AF358619

Evolutionary Algorithms in Intelligent Systems

Evolutionary Algorithms in Intelligent Systems

Editors

Alfredo Milani
Arturo Carpi
Valentina Poggioni

MDPI • Basel • Beijing • Wuhan • Barcelona • Belgrade • Manchester • Tokyo • Cluj • Tianjin

Editors
Alfredo Milani
University of Perugia
Italy

Arturo Carpi
University of Perugia
Italy

Valentina Poggioni
University of Perugia
Italy

Editorial Office
MDPI
St. Alban-Anlage 66
4052 Basel, Switzerland

This is a reprint of articles from the Special Issue published online in the open access journal *Mathematics* (ISSN 2227-7390) (available at: https://www.mdpi.com/journal/mathematics/special_issues/Mathematics_Evolutionary_Algorithms).

For citation purposes, cite each article independently as indicated on the article page online and as indicated below:

LastName, A.A.; LastName, B.B.; LastName, C.C. Article Title. *Journal Name* **Year**, *Article Number*, Page Range.

ISBN 978-3-03943-611-8 (Hbk)
ISBN 978-3-03943-612-5 (PDF)

Contents

About the Editors

Alfredo Milani (Professor) is an Associate Professor at the Department of Mathematics and Computer Science, University of Perugia, Italy. He received his doctoral degree in Information Science from the University of Pisa, Italy. His research interests include the broad area of artificial intelligence with a focus on evolutionary algorithms, and applications to planning, user interfaces, e-learning, and web-based adaptive systems. He is the author of many international journal papers and chair of international conferences and workshops. He is the scientific leader of the KitLab research lab at the University of Perugia.

Arturo Carpi is a full professor at the Univesity of Perugia, Italy. His research interests include theoretical aspects of coding, languages, and context-free grammar, on which he has authored several relevant articles. Recently, he has also been active in the area of memetic particle swarm evolution.

Valentina Poggioni (Doctor) is an Assistant Professor and Researcher in Artificial Intelligence at the University of Perugia in Italy, where she teaches Machine Learning, Data Mining, and Artificial Intelligence. She attained her Ph.D. at the University "Roma Tre" in Rome, where she studied automated planning and temporal and multivalued logics. Currently, her main interests are in evolutionary computation and machine learning with a particular focus on neural networks, and the application of evolutionary algorithms to neural network optimization.

 *mathematics*

Editorial

Evolutionary Algorithms in Intelligent Systems

Alfredo Milani

Department of Mathematics and Computer Science, University of Perugia, 06123 Perugia, Italy; milani@unipg.it

Received: 17 August 2020; Accepted: 29 August 2020; Published: 10 October 2020

Evolutionary algorithms and metaheuristics are widely used to provide efficient and effective approximate solutions to computationally difficult optimization problems. Successful early applications of the evolutionary computational approach can be found in the field of numerical optimization, while they have now become pervasive in applications for planning, scheduling, transportation and logistics, vehicle routing, packing problems, etc. With the widespread use of intelligent systems in recent years, evolutionary algorithms have been applied, beyond classical optimization problems, as components of intelligent systems for supporting tasks and decisions in the fields of machine vision, natural language processing, parameter optimization for neural networks, and feature selection in machine learning systems. Moreover, they are also applied in areas like complex network dynamics, evolution and trend detection in social networks, emergent behavior in multi-agent systems, and adaptive evolutionary user interfaces to mention a few. In these systems, the evolutionary components are integrated into the overall architecture and they provide services to the specific algorithmic solutions. This paper selection aims to provide a broad view of the role of evolutionary algorithms and metaheuristics in artificial intelligent systems.

A first relevant issue discussed in the volume is the role of multi-objective meta-optimization of evolutionary algorithms (EA) in continuous domains. The challenging tasks of EA parameter tuning are the many different details that affect EA performance, such as the properties of the fitness function as well as time and computational constraints. EA meta-optimization methods in which a metaheuristic is used to tune the parameters of another (lower-level) metaheuristic, which optimizes a given target function, most often rely on the optimization of a single property of the lower-level method. A multi-objective genetic algorithm can be used to tune an EA, not only to find good parameter sets considering more objectives at the same time but also to derive generalizable results that can provide guidelines for designing EA-based applications. In a general framework for multi-objective meta-optimization, it is necessary to show that "going multi-objective" allows one to generate configurations, besides optimally fitting an EA.

A significant example of this approach is the application of differential evolution-based methods for the optimization of neural networks (NN) structure and NN parameter optimization. Such an adaptive differential evolution system can be seen as an optimizer which applies mutation and crossover operators to vary the structure of the neural network according to per layer strategies. Self-adaptive variants of differential evolution algorithms tune their parameters on the go by learning from the search history. Adaptive differential evolution with an optional external archive and self-adaptive differential evolution are well-known self-adaptive versions of differential evolution (DE). They are optimization algorithms based on unconstrained search.

Another relevant general area of evolutionary algorithms is represented by the Particle Swarm Optimization (PSO), which is based on the concept of swarm of particles, i.e., individual solutions and computational entities. A swarm extends the concept of a set of solutions of the early classical genetic algorithms to a set of related, coordinated, and interacting search threads. On this basis, it is interesting to explore the many variants of PSO, like, for instance, the memetic variant. The memetic evolution of local search operators can be introduced in PSO continuous/discrete hybrid search spaces. The evolution of local search operators overcome the rigidity of uniform local search strategies.

The memes provide each particle of a PSO scheme with the ability to adapt its exploration dynamics to the local characteristics of the search space landscape. A further step is to apply a co-evolving scheme to PSO. Co-evolving memetic PSO can evolve both the solutions and their associated memes, i.e., the local search operators.

PSO can be straightforwardly adapted to multi-objective optimization, an innovative contribution, explore methods for obtaining high convergence and uniform distributions, which remains a major challenge in most metaheuristic multi-objective optimization problems. The selected article proposes a novel multi-objective PSO algorithm based on the Gaussian mutation and an improved learning strategy to improve the uniformity of external archives and current populations.

A common trend of an evolutionary algorithm scenario is the constantly increasing number of new proposals of nature-inspired metaheuristics. These proposed approaches usually take inspiration from groups of distributed agents existing in nature, i.e., ants, flock of birds, bees, etc., which apply simple local rules, but globally result in a complex emerging behavior which optimizes some specific feature, i.e., amount of found food, shortest path, the change of surviving to predators, etc. We found it to be interesting to propose, among the selected articles, an application to Internet of Things, in particular, the optimal task allocation problem in wireless sensor networks. The nature-inspired evolutionary algorithm proposed for wireless sensor networks is the recently developed Social Network Optimization, which is a significant example of using behavioral rules of social network users for obtaining an emerging optimization behavior.

Acknowledgments: This work has been partially supported by the Italian Ministry of Research under PRIN Project "PHRAME" Grant n.20178XXKFY.

Conflicts of Interest: The author declares no conflict of interest.

 *mathematics*

Article

What Can We Learn from Multi-Objective Meta-Optimization of Evolutionary Algorithms in Continuous Domains?

Roberto Ugolotti [1], Laura Sani [2] and Stefano Cagnoni [2,*]

[1] Camlin Italy, via Budellungo 2, 43123 Parma, Italy; r.ugolotti@camlintechnologies.com
[2] Department of Engineering and Architecture, University of Parma, Parco Area delle Scienze 181/a, 43124 Parma, Italy; laura.sani@unipr.it
*   Correspondence: stefano.cagnoni@unipr.it; Tel.: +39-05-2190-5731

Received: 7 January 2019; Accepted: 28 February 2019; Published: 4 March 2019

Abstract: Properly configuring Evolutionary Algorithms (EAs) is a challenging task made difficult by many different details that affect EAs' performance, such as the properties of the fitness function, time and computational constraints, and many others. EAs' meta-optimization methods, in which a metaheuristic is used to tune the parameters of another (lower-level) metaheuristic which optimizes a given target function, most often rely on the optimization of a single property of the lower-level method. In this paper, we show that by using a multi-objective genetic algorithm to tune an EA, it is possible not only to find good parameter sets considering more objectives at the same time but also to derive generalizable results which can provide guidelines for designing EA-based applications. In particular, we present a general framework for multi-objective meta-optimization, to show that "going multi-objective" allows one to generate configurations that, besides optimally fitting an EA to a given problem, also perform well on previously unseen ones.

Keywords: evolutionary algorithms; multi-objective optimization; parameter puning; parameter analysis; particle swarm optimization; differential evolution; global continuous optimization

---

1. Introduction

This paper investigates Evolutionary Algorithms (EAs) tuning from a multi-objective perspective. In particular, a set of experiments exemplify some of the relevant additional hints that a general multi-objective EA-tuning (Meta-EA) environment can provide, regarding the impact of EAs' parameters on EAs' performance, with respect to the single-objective EA-tuning environment of which it is a very simple extension.

Evolutionary Algorithms [1] have been very successful in solving hard, multi-modal, multi-dimensional problems in many different tasks. Nevertheless, configuring EAs is not simple and implies critical decisions that are taken based, as summarized below, on a number of factors, such as: (i) the nature of the problem(s) under consideration, (ii) the problem's constraints, such as the restrictions imposed by computation time requirements, (iii) an algorithm's ability to generalize results over different problems, and (iv) the quality indices used to assess its performance.

Problem Features

When dealing with black-box real-world problems it is not always easy to identify the mathematical and computational properties of the corresponding fitness functions (such as modality, ruggedness, isotropy of the fitness landscape, see [2]). Because of this, EAs are often applied acritically, using "standard" parameter settings which work reasonably on most problems but most often lead to sub-optimal solutions.

Generalization

An algorithm that effectively optimizes a certain function should optimize as effectively functions characterized by the same computational properties. An interesting study on this issue is the investigation of "algorithm footprints" [3].

Some configurations of EAs, among which "standard" settings are usually comprised, can reach similar results on many problems, while others may exhibit performance characterized by a larger variability. While it is obviously important to find a good parameter set for a specific EA dealing with a specific problem, it is even more important to understand how much changing it can affect the performance of the EA.

Constraints and Quality Indices

Comparing algorithms (or different instances of the same algorithm) requires a precise definition of the conditions under which the comparison is made. As will be shown later in the plots Q_{10K} and Q_{100K} in Figure 7 (top left), convergence to a good solution can occur with very different modalities. Some parameter settings may lead to fast convergence to a sub-optimal solution, while others may need many more fitness evaluations to converge, but lead to better solutions. In several real-world applications it is often sufficient to reach a point which is "close enough" to the global optimum; in such cases, an EA that is consistently able to reach good sub-optimal results timely is to be preferred to slower, although more precise, algorithms. Instead, in problems with weaker time constraints, an EA that keeps refining the solution over time, even very slowly, is usually preferable.

The previous considerations indicate that comparing different algorithms is very difficult because, for the comparison to be fair, each algorithm should be used "at its best" for the given problem. In fact, there are many examples in the literature where the effort spent by the authors on tuning and optimizing the method they propose is much larger than the effort spent on tuning the ones to which it is compared. This may easily lead to biased interpretations of the results and to wrong conclusions.

The importance of methods (usually termed Meta-EAs) that tune EAs' parameters to optimize their performance has been highlighted since 1978 [4]. However, mainly due to the relevant computational effort they require, Meta-EAs and other parameter tuning techniques have become a mainstream research topic only recently.

We are aware that using as Meta-EA an algorithm whose behavior, as well, depends on its setup, would imply that the Meta-EA itself should undergo parameter tuning. There are obvious practical reasons related to the method's computational burden for not doing so. As well, it can be argued that if the application of a Meta-EA can effectively lead to solutions that are closer to the global optimum for the problem at hand than those found by a standard setting of the algorithm that is being tuned, then, even supposing one uses several optimization meta-levels, the improvement margins for each higher-level Meta-EA become smaller and smaller with the level. This intuitively implies that the variability of the results depending on the higher-level Meta-EAs parameter settings also becomes smaller and smaller with the level. Therefore, even if, most probably, better settings of the Meta-EA could further improve the optimization performance, we consider that a "standard" setting of the Meta-EA is generally enough to achieve some relevant performance improvement with respect to a random setting.

In [5], we proposed SEPaT (Simple Evolutionary Parameter Tuning), a single-objective Meta-EA in which GPU-based versions of Differential Evolution (DE, [6]) and Particle Swarm Optimization (PSO, [7]) were used to tune PSO on some benchmark functions, obtaining parameter sets that yielded results comparable with the state of the art and better than "standard" or manual settings.

Even if results were good, the approach was mainly practical, aimed at providing one set of good parameters, but no hints about their generality or about the reasons why they had been selected. One of the main limitations of the approach was related to its performing a single-objective optimization, which prevented it from considering other critical goals, such as generalization, besides the obvious one to optimize an EA's performance on a given problem.

In this paper, we go far beyond such results, investigating what additional hints a multi-objective approach can provide. To do so, we use a very general framework, which we called EMOPaT (Evolutionary Multi-Objective Parameter Tuning), that was described in [8]. EMOPaT uses the well-known Multi-Objective Evolutionary Algorithm (MOEA) Non-dominated Sorting Genetic Algorithm (NSGA-II, [9]) to automatically find good parameter sets for EAs.

The goal of this paper is not proposing EMOPaT as a reference environment. Instead, we use it, as virtually the simplest possible multi-objective derivation of SEPaT, to focus on some of the many additional hints that a multi-objective approach to EA tuning can provide with respect to a single-objective one. We are well conscious that more sophisticated and possibly better performing environments aimed at the same goal can be designed. SEPaT and EMOPaT have been developed with no intent to advance the state of the art of meta-optimization algorithms but as generic frameworks, with as few specific features as possible, aimed at studying EA meta-optimization. Consistently with this principle, within EMOPaT, we use NSGA-II as the multi-objective algorithm tuner, since it is possibly the most widely available, generally well-performing and easy to implement multi-objective stochastic optimization algorithm. Indeed, NSGA-II can be considered a natural extension of a single-objective genetic algorithm (GA) to multi-objective optimization. As well, we chose to test EMOPaT in tuning PSO and DE for no other reasons than the easy availability and good computational efficiency of these algorithms. EMOPaT is a general environment and can be used to tune virtually any other EA or metaheuristic.

EMOPaT is not only aimed at finding parameter sets that achieve good results considering the nature of the problems, the quality indices and, more in general, the conditions under which the EA is tuned. It allows one to extract information about the parameters' semantics and the way they affect the algorithm by analyzing the Pareto fronts approximated by the solutions obtained by NSGA-II. A similar strategy has been presented by [10] under the name of *innovization*(innovation through optimization).

As well, we show that EMOPaT can evolve parameter sets that let an algorithm perform well not only on the problem(s) on which it has been tuned, but also on others. Section 2 briefly introduces the three EAs used in our experiments, Section 3 reviews the methods that inspired our work, and Section 4 describes EMOPaT. In Section 5 we first use EMOPaT to find good parameter sets for optimizing the same function under different conditions: doing so, we show that the analysis of EMOPaT's results can clarify the role of EAs' parameters and study EMOPaT's generalization abilities; finally, EMOPaT is used to optimize seven benchmark functions and generalize its results to previously unseen functions. Section 6 summarizes all results and suggests possible future extensions of this work.

Additionally, in a separate appendix, we demonstrate that EMOPaT can be considered an extension of SEPaT and has equivalent performance in solving single-objective problems, as well as assessing its correct behavior by considering some controlled situations, on which we show it to be able to perform tuning as expected.

2. Background

2.1. Differential Evolution

In every generation of DE, each individual in the population acts as a parent vector for which a donor vector $\vec{D}_i$ is created. A donor vector is generated by combining three random and distinct individuals $\vec{X}_{r1}$, $\vec{X}_{r2}$ and $\vec{X}_{r3}$ according to this simple mutation equation:

$$\vec{D}_i = \vec{X}_{r1} + F \cdot (\vec{X}_{r2} - \vec{X}_{r3}) \tag{1}$$

where F (scale factor) is usually in the interval [0.4, 1]. Several different mutation strategies have been applied to DE; in our work, along with the *random* mutation reported above, we consider *best* and *target-to-best* (or *TTB*) mutation strategies, whose definitions are, respectively:

$$\vec{D}_i = \vec{X_{best}} + F \cdot (\vec{X_{r1}} - \vec{X_{r2}}) \tag{2}$$

$$\vec{D}_i = \vec{X}_i + F \cdot (\vec{X_{best}} - \vec{X}_i) + F \cdot (\vec{X_{r1}} - \vec{X_{r2}}) \tag{3}$$

After mutation, every parent-donor pair generates a child ($\vec{T}_i$), called trial vector, by means of a crossover operation. Two kinds of crossover are usually employed in DE: *binomial* and *exponential* (see [11] for more details). Both crossover strategies depend on the crossover rate CR. The newly generated individual $\vec{T}_i$ is evaluated by comparing its fitness to its parent's. The better individual survives and will be part of the next generation.

2.2. Particle Swarm Optimization

In PSO ([7]), a set of particles moves within the search space, according to these equations, that describe particle i's velocity and position:

$$\vec{v_i}(t) = w \cdot v_i(\vec{t-1}) + c_1 \cdot rand() \cdot (\vec{BP}_i - P_i(\vec{t-1})) + c_2 \cdot rand() \cdot (B\vec{G}P_i - P_i(\vec{t-1})) \tag{4}$$

$$\vec{P_i}(t) = P_i(\vec{t-1}) + \vec{v_i}(t) \tag{5}$$

where c_1, c_2, and w (inertia factor) are real-valued constants, $rand()$ returns random values uniformly distributed in $[0,1]$, $\vec{BP}_i$ is the best-fitness position visited so far by the particle, and $B\vec{G}P_i$ the best-fitness position visited so far by any individual in the particle's neighborhood, that can comprise the entire swarm or only a subset. In this work, we consider three of the most commonly used neighborhood topologies (see Figure 1).

Figure 1. The three PSO topologies used in this work: global, ring, and star.

2.3. NSGA-II

The NSGA-II algorithm is basically a classical GA in which selection is based on the so-called non-dominated sorting. In case two individuals have the same rank, the one with the greater crowding distance is selected. This distance can take into consideration the fitness values or the encoding of the individuals, to increase the diversity of the results or of the population, respectively. In this work, NSGA-II crossover and mutation rates have been set as suggested in [9], while we have set the population size and the number of generations "manually", based on the complexity of the problem at hand.

3. Related Work

The importance of parameter tuning has been frequently addressed in the last years, not only in theoretical or review papers such as [12] but also in papers with extensive experimental evidence which provide a critical assessment of such methods. In [13], while recognizing the importance of finding a good set of parameters, the authors even suggest that using approaches to algorithm tuning that are computationally demanding may be almost useless, since a relatively limited random search in the algorithm parameter space can often offer good results.

Meta-optimization algorithms can be grouped into two main classes:

- Parameter tuning: the parameters are chosen offline and their values do not change during evolution, which is the case of interest for this paper;
- Parameter control [14]: the parameters may vary during evolution, according to a strategy that depends on the results that are being achieved. These changes are usually driven either by fitness improvement (or by its lack) or by properties of the evolving population, like diversity or entropy.

Along with Meta-EAs, several methods which do not strictly belong to that class but use similar paradigms have been proposed: one of the most successful is Relevance Estimation and Value Calibration (REVAC) by [15], a method inspired by the Estimation of Distribution Algorithm (EDA, [16]) that was able ([17]) to find parameter sets that improved the performance of the winner of the competition on the CEC 2005 test-suite [18]. In [19], PSO tuned itself to optimize neural network training; Reference [20] used a simple metaheuristic, called Local Unimodal Sampling, to tune DE and PSO, obtaining good performance while discovering unexpectedly good parameter settings. Reference [21] proposed ParamILS, whose local search starts from a default parameter configuration which is then iteratively improved by modifying one parameter at a time. Reference [22] used a Meta-EA as an optimization method in a massively parallel system to generate on-the-fly optimizers that directly solved the problem under consideration. In [23], the authors propose a self-adaptive DE for feature selection.

Other approaches to parameter tuning include model-based methods like Sequential Parameter Optimization (SPO) proposed by [24] and racing algorithms [25,26]: they generate a population of possible configurations based on a particular distribution; members of this population are then tested and possibly discarded as soon as a statistical test shows that there is at least another individual which outclasses them; these operations are repeated until a set of good configurations is obtained. A recent trend approaches parameter tuning as a two-level optimization problem [27,28].

The first multi-objective Meta-EA was proposed in [29] where NSGA-II was used to optimize speed and precision of four different algorithms. However, that work took into consideration only one parameter at a time, so the approach described therein cannot be considered a full parameter set optimization algorithm. A similar method has been proposed by [30]. The authors describe a variation of a MOEA called Multi-Function Evolutionary Tuning Algorithm (M-FETA), in which the performance of a GA on two different functions represent the different goals that the MOEA must optimize; the final goal is to discriminate algorithms that perform well on a single function from those that do on more than one, respectively called "specialists" and "generalists", following the terminology introduced by [31].

In [32], the authors propose an interesting technique, aimed at identifying the best parameter settings for different possible computational budgets (i.e., number of fitness evaluations) up to a maximum. This is obtained using a MOEA in which the fitness of an individual is a vector whose components are the fitness values obtained in every generation. In this way, it is possible to find a family of parameter sets which obtain the best results with different computational budgets.

A comprehensive review of Meta-EAs can be found in [33].

More recently, MO-ParamILS has been proposed as a multi-objective extension of the state-of-the-art single-objective algorithm configuration framework ParamILS [34]. This automatic algorithm produces good results on several challenging bi-objective algorithm configuration scenarios. In [35], MO-ParamILS is used to automatically configure a multi-objective optimization algorithm in a multi-objective fashion.

4. EMOPaT, a General Framework for Multi-Objective Meta-Optimization

This section describes EMOPaT's main structure and operation, introduced in [5] as a straightforward multi-objective extension of the corresponding single-objective general framework SEPaT.

SEPaT and EMOPaT share the same very general scheme, presented in Figure 2.

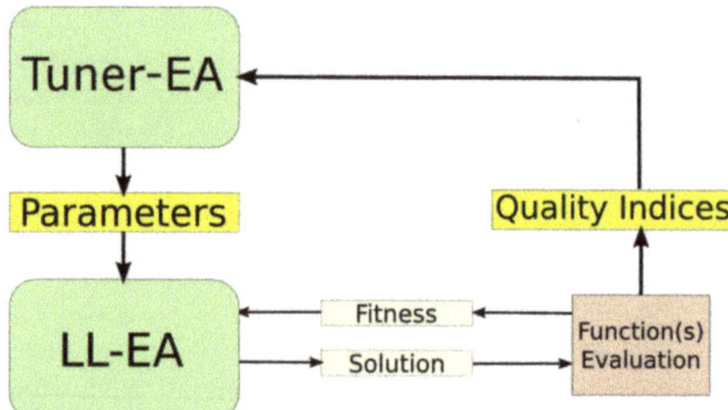

Figure 2. Scheme of SEPaT/EMOPaT. The lower part represents a classical EA. In the meta-optimization process, each individual of Tuner-EA represents a set of Parameters. For each set, the corresponding instance of the lower-level EA (LL-EA) is run N times to optimize the objective function(s). Quality indices (one for SEPaT, more than one for EMOPaT) are values that provide a global evaluation of the results obtained by LL-EA in these runs.

The block in the lower part of the image represents a traditional optimization problem in which an EA, referred to as Lower-Level EA (LL-EA) optimizes one or more objective functions. The Tuner EA operates within the search space of the parameters of the LL-EA. This means that the tuner evolves a population of possible parameter sets of LL-EA parameters. Each parameter set corresponds to an instance of LL-EA that is tested N times on LL-EA's objective function(s) (from now on, we will consider "configuration" and "parameter set" as equivalent terms). The N results are synthesized into one or more "Quality Indices" that represent the objective function(s) of the tuner.

The difference between SEPaT and EMOPaT therefore stands in the different number of quality indices. In SEPaT, any single-objective EA can be used as Tuner EA, while EMOPaT requires a multi-objective EA. In the case described in this paper, we used NSGA-II.

It should be noticed that as evidenced in the figure, the tuning of the (usually, but not necessarily, single-objective) LL-EA may be aimed at finding the best "generalist" setting for optimizing any number of functions. For instance, in [5] PSO and DE were used as tuners in SEPaT to optimize the behavior of PSO over 8 objective functions. In that case, an EA configuration was considered better than another if it obtained better results over the majority of the functions. The quality index, in this case, was therefore a score computed according to a tournament-like comparison among the individuals.

In [5], the parameter set found by SEPaT was compared to the set found using *irace* [25,36] and to "standard" parameters, with results similar to *irace* and better than the "standard" settings.

On the one hand, using this approach, besides allowing one to synthesize the results as a single score, brings the advantage that the functions for which the LL-EAs are tuned do not need to assume values within comparable ranges, avoiding the need for normalization. On the other hand, being based on a comparison may sometimes limit the effectiveness of this approach. In fact, a configuration may win even if it cannot obtain good results on some of the functions, since it is required only to perform better than the others on the majority of them. Therefore, the resulting parameter sets, despite being good on average, may not be as good on all functions. This is one of the limitations that EMOPaT tries to overcome (see Section 5.2).

The multiple objectives taken into consideration by EMOPaT may differ depending on the function under consideration, the quality index considered, or the constraints applied, such as the number of evaluations, time constraints or others. The output of the tuning process is not a single solution as in SEPaT, but an entire set of non-dominated EA configurations, i.e., ideally, a sampling of the Pareto front for the objectives under consideration (see Figure 6 for two examples of Pareto fronts, highlighted in yellow). This allows a developer to analyze the parameters' selection strategy more in depth. We think that this approach can be particularly relevant, in light of the conclusions drawn in [37]: according to the outcome of the experiments, even if the Meta-EAs they considered performed better than SPO and REVAC, the authors pointed out that they were unable to provide insights about EA parameters.

Parameter Representation

Since the tuner algorithms we consider are real-valued optimization methods, we need a proper representation of the nominal parameters of the LL-EA, i.e., the parameters that encode choices among a limited set of options. We opted for representing each nominal parameter as a real-valued vector with as many elements (genes) as the options available: the actual choice is the one that corresponds to the gene with the largest value. For instance, if the parameter to optimize is PSO topology, we can choose between *ring*, *star* and *global* topology. Each individual in the tuner represents this setting as a three-dimensional vector whose largest element determines the topology used in the LL-EA configuration. These particular genes are mutated and crossed-over following NSGA-II rules just like any other. Figure 3 shows how DE and PSO configurations are encoded.

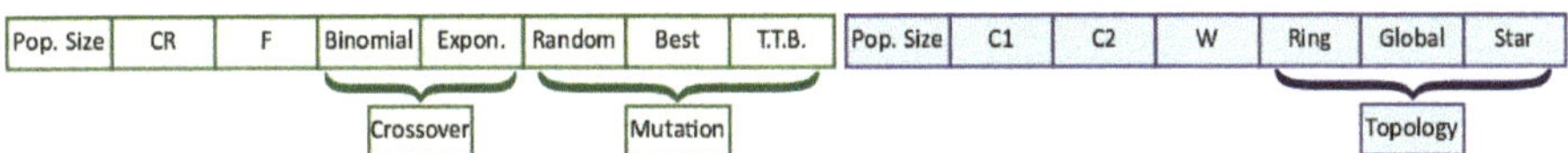

Figure 3. Encoding of DE (left) and PSO (right) configurations in a tuner EA.

5. Experimental Evaluation

In this section, we discuss the results of some experiments in which we optimize different performance criteria that can assess the effectiveness of an EA in solving a given optimization task. We take into consideration "classical" criteria pairs, such as solution quality vs. convergence speed, as well as experiments in which the different criteria are represented by different constraints on the available resources (e.g., different fitness evaluation budgets).

To do so, we use DE and PSO as LL-EAs and NSGA-II as EMOPaT's Tuner-EA. Table 1 shows the ranges within which we let PSO and DE parameters change in our tests. During the execution of the Tuner-EA, all values are actually normalized in the range $[0, 1]$; a linear scale transformation is then performed whenever a LL-EA is instantiated.

Table 1. Search ranges for the DE and PSO parameters. We chose ranges that are wider than those usually considered in the literature, to allow SEPaT and EMOPaT to "think outside the box", and possibly find unusual parameter sets.

Differential Evolution		Particle Swarm Optimization	
Population Size	$[4, 300]$	Population Size	$[4, 300]$
Crossover Rate (CR)	$[0.0, 1.0]$	Inertia Factor (w)	$[-0.5, 1.5]$
Scale Factor (F)	$[0.0, 2.0]$	c_2	$[-0.5, 4.0]$
Crossover	{*binomial, exponential*}	c_1	$[-0.5, 4.0]$
Mutation	{*random, best, target-to-best*}	Topology	{*ring, star, global*}

The computation load of the meta-optimization process is heavily dependent on the cost of a single optimization process. If we term t the average time needed for a single run of the LL-EA (which corresponds, for the Tuner-EA, to one fitness evaluation), then the upper bound for the time T needed for the whole process is:

$$T = t \cdot \texttt{<Tuner generations>} \cdot \texttt{<Tuner population size>} \cdot N \qquad (6)$$

since we can consider the computation time requested by the tuner's search operators to be negligible with respect to a fitness evaluation. This process can be highly parallelized, since all N repetitions, as well as all evaluations of a population can be run in parallel if enough resources are available. In our tests, we used an 8-core 64-bit Intel(R) CoreTM i7 CPU running at 3.40 GHz; we chose not to parallelize the optimization process but we preferred to parallelize independent runs of the tuners.

EMOPaT has been tested on some functions from the CEC 2013 benchmark [38], with the only difference that the function minima were set to 0.

The code used to perform the tests is available online at http://ibislab.ce.unipr.it/software/emopat.

5.1. Multi-Objective Single-Function Optimization Under Different Constraints

A multi-objective experiment can optimize different functions, the same function under different conditions, etc. Thus, optimizing a single function under different constraints can be seen as a particular case of multi-objective optimization. In this section, we report the results of tests on single-function optimization under different fitness evaluations budgets. Similar experiments can be performed evaluating the function under different conditions (e.g., different problem dimensions) or according to different quality indices as we did in [8] where we considered two objectives (fitness and fitness evaluations budget) for a single function. With respect to that work, the main additional contribution of this section is showing how EMOPaT can be used to generalize the behavior of an EA in optimizing a function when working under different conditions. We consider the following set of quality indices:

$\{Q_{X_i}\} \triangleq$ best results after $\{X_i\}$ fitness evaluations, averaged over N runs.

We performed four different tests considering, in each of them, one of the four functions shown in Table 2. Our objectives were the best-fitness values reached after $1000, 10,000$ and $100,000$ function evaluations, namely Q_{1K}, Q_{10K}, Q_{100K}. Each test was run 10 times. Doing so, we expected we would favor the emergence of patterns related with the impact of a parameter when looking for "fast-converging" or "slow-converging" configurations. Table 2 summarizes the experimental setup for these experiments.

Firstly, we analyze the LL-EA parameter sets evolved under the different criteria. To do so, we merge the populations of the ten independent runs and, from this pool, we select, for each objective, the top 10% of the best solutions. For most parameters there is a clear trend as their values monotonically grow or decrease as the fitness evaluations budget increases (see Table 3). This result suggests that, in these cases, it may not be necessary to keep track of all possible computational budgets as in [32], but that the optimal parameters for intermediate objectives may be inferred by interpolating the ones found for the objectives actually taken into consideration; consequently, a developer can use this information to tune its algorithm according to his own budget constraints. Nevertheless, while this can be true for a function, such trends are rarely consistent through different functions, preventing one from drawing more general conclusions.

Table 2. Single-function optimization with different fitness evaluation budgets. Experimental settings.

EMOPaT settings
Population Size = 64, 60 Generations, Mutation Rate = 0.125, Crossover Rate = 0.9
Function settings
30-dimensional Sphere, Rastrigin, Rosenbrock, Griewank (one for each experiment) evaluation criterion = best fitnesses after 1000, 10,000, 100,000 evaluations averaged over $N = 15$ repetitions.

Let us analyze in more details the results on the Rastrigin and the Sphere functions (similar conclusions can be drawn for the other functions). Figures 4 and 5 show the boxplots of some parameters for the top 10% DE and PSO configurations on these two functions. When parameters are nominal, a bar chart plots the selection frequencies for each option. For the Sphere function, the boxplots of Q_{10K} and Q_{100K} are very similar, pointing out that for this function, 10,000 evaluations are usually sufficient to reach convergence.

Table 3. Trends of DE and PSO parameter values versus fitness evaluations budget. Upward arrows denote parameter values increasing with the number of evaluations, downward arrows the opposite. A dash denotes no clear trend. For nominal parameters, the table reports the most frequently selected choice. If the choice changes within the top solutions for different evaluation budgets, an arrow shows the direction of this change as the budget increases.

Differential Evolution					
Function	**PopSize**	**CR**	**F**	**Mutation**	**Crossover**
Sphere	↘	↘	↗	*target-to-best*	*binomial→exponential*
Rastrigin	↗	↗	↗	*—*	*binomial→exponential*
Griewank	↗	↘	↘	*target-to-best→random*	*binomial*
Rosenbrock	↗	↗	↗	*target-to-best→best,random*	*binomial→exponential*

Particle Swarm Optimization					
Function	**PopSize**	**w**	c_1	c_2	**Topology**
Sphere	↗	↘	—	↗	*ring*
Rastrigin	↗	↘	↗	↘	*global*
Griewank	↗	↘	↗	↗	*ring*
Rosenbrock	↗	↗	↘	↗	*global→ring*

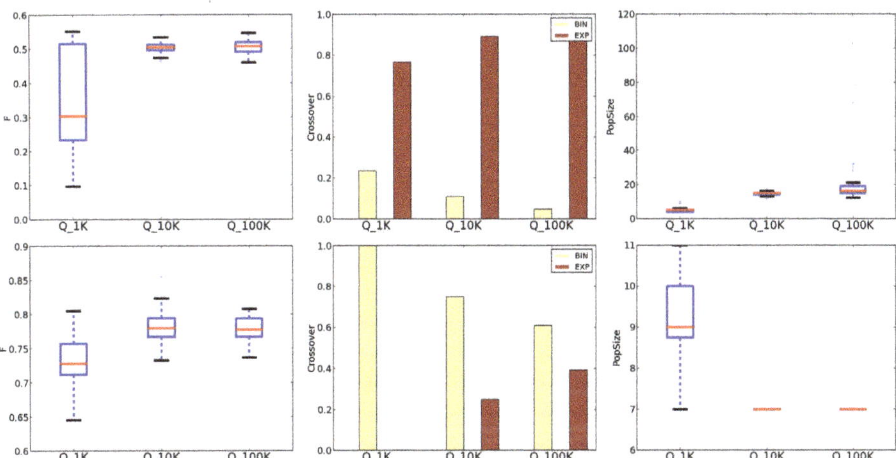

Figure 4. DE parameters of the top solutions (best 10%) for the 30-dimensional Rastrigin (first row) and Sphere (second row) functions, with an available budget of 1000, 10,000, and 100,000 fitness evaluations. Bar plots indicate the normalized selection frequency. Descending/ascending trends for all parameter values are clearly visible.

To evaluate the hypothesis that intermediate budgets can be inferred from the results obtained on the objectives that have actually been optimized, one can generate new solutions in two ways:

- infer them as the mean of two top solutions found for two different objectives between which the new objective lies. This approach presents some limitations: (i) the parameter must have a clear trend, (ii) some policy for nominal parameters must be defined if the two reference configurations have different settings, (iii) there is no guarantee that all intermediate values of a parameter correspond to valid configurations of the algorithm;
- select them from the Pareto front by plotting the set of all the individuals obtained at the end of ten independent runs on the two objectives of interest, estimate the Pareto front based on those solutions and randomly pick one configuration that lies on it, in an intermediate position between the extremes (see an example related with the Rastrigin function in Figure 6).

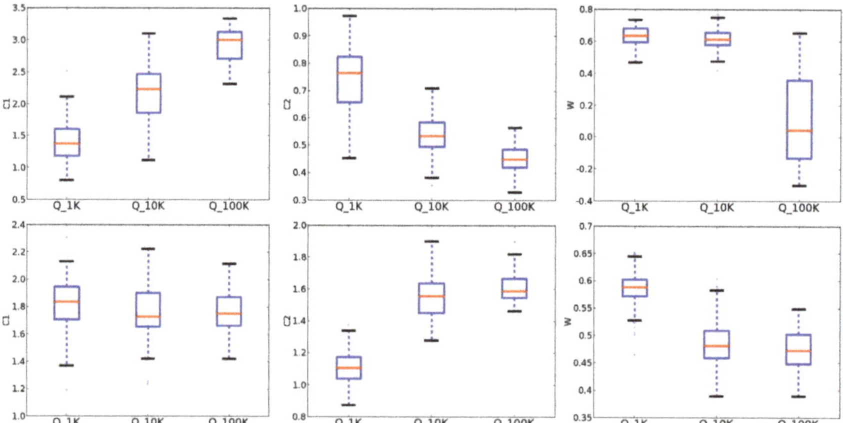

Figure 5. PSO parameters of the top solutions (best 10% of the population) for the 30-dimensional Rastrigin (first row) and Sphere (second row) functions, with an available budget of 1000, 10,000, and 100,000 fitness evaluations. Descending/ascending trends for all parameter values are clearly visible.

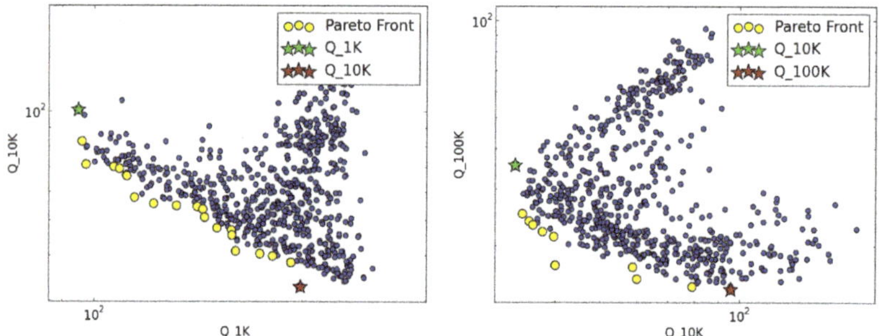

Figure 6. Fitness values of all the solutions found in ten independent runs of EMOPaT for the three criteria, plotted pairwise for adjacent values of the budget. The green and red stars represent the Top Solutions for each objective, yellow circles are candidate solutions for intermediate evaluation budgets.

Table 4 shows the parameters of the best solutions found for the Rastrigin and Sphere functions for the three objectives and of four intermediate solutions generated by the two methods. The ones indicated by A lie between Q_{1K} and Q_{10K} and the ones indicated by B between Q_{10K} and Q_{100K}. It can be noticed that the values of the parameter sets generated using the Pareto Front differ from both the ones inferred as a weighted mean of neighboring ones and the top solutions. In some cases (as with DE on Sphere) these solutions use a mutation type that is never considered by the top solutions. For the nominal parameters of the inferred solutions, we chose to consider both options when the two top solutions disagreed, distinguishing the two solutions by an index (e.g., A_1 and A_2).

Figure 7 shows the performance of the configurations considered in Table 4, averaged over 100 independent runs, for DE and PSO. The solid lines represent the Top Solutions; as expected, after 1000 evaluations (see the plots on the right) Q_{1K} is the best-performing configuration, while Q_{100K} is slower in the beginning but is the best at the end of the evolution. In most cases, the inferred solutions have a performance that lies between the performance of the two top solutions used as starting points. The results obtained on the Rastrigin function (first row of Figure 7) are particularly clear: in the first 1000 evaluations, Q_{1K} performs best, then it is surpassed by Inferred A, followed by Q_{10K}, Pareto B, and finally Q_{100K}; this example seems to confirm our general hypothesis. A relevant exception is A_2 in DE Sphere (Figure 7, last row) that performs worse than all others: since its only difference with A_1 is the crossover type, this suggests that it is not possible to infer nominal parameters reliably unless one is clearly prevalent.

Table 4. DE and PSO configurations for the same objective function with three different fitness evaluations budgets."Top Solutions" are the best-performing sets on each objective; "Inferred" refers to the ones obtained averaging Top Solutions; "From Pareto" are extracted from the Pareto front obtained considering the objectives pairwise (see Figure 6).

	Differential Evolution						
	Method	**Configuration**	**PopSize**	**CR**	**F**	**Mutation**	**Crossover**
Rastrigin	Top Solutions	Q_{1K}	9	0.214	0.736	*target-to-best*	*binomial*
		Q_{10K}	7	0.053	0.754	*target-to-best*	*exponential*
		Q_{100K}	7	0.039	0.784	*target-to-best*	*exponential*
	Inferred	A_1, A_2	8	0.134	0.745	*target-to-best*	*bin., exp.*
		B	7	0.046	0.769	*target-to-best*	*exponential*
	From Pareto	A	9	0.217	0.763	*target-to-best*	*binomial*
		B	7	0.006	0.769	*target-to-best*	*binomial*
Sphere	Top Solutions	Q_{1K}	5	0.022	0.229	*random*	*binomial*
		Q_{10K}	14	0.363	0.508	*random*	*exponential*
		Q_{100K}	30	0.043	0.521	*random*	*exponential*
	Inferred	A_1, A_2	9	0.192	0.368	*random*	*bin., exp.*
		B	22	0.203	0.514	*random*	*exponential*
	From Pareto	A	8	0.023	0.520	*target-to-best*	*exponential*
		B	15	$0.50E-3$	0.498	*target-to-best*	*binomial*
	Particle Swarm Optimization						
	Method	**Configuration**	**PopSize**	**w**	c_1	c_2	**Topology**
Rastrigin	Top Solutions	Q_{1K}	18	0.560	1.195	0.789	*global*
		Q_{10K}	26	0.579	2.492	0.671	*global*
		Q_{100K}	76	-0.251	2.533	0.487	*global*
	Inferred	A	22	0.569	1.844	0.730	*global*
		B	51	0.164	2.513	0.579	*global*
	From Pareto	A	27	0.678	0.949	0.587	*global*
		B	60	0.297	3.132	0.481	*global*
Sphere	Top Solutions	Q_{1K}	11	0.603	1.882	1.105	*ring*
		Q_{10K}	14	0.510	1.998	1.483	*ring*
		Q_{100K}	15	0.449	1.725	1.667	*ring*
	Inferred	A	12	0.557	1.940	1.294	*ring*
		B	15	0.480	1.861	1.575	*ring*
	From Pareto	A	14	0.649	1.764	1.082	*ring*
		B	15	0.480	1.681	1.635	*ring*

Finally, to compare EMOPaT with a state-of-the-art tuner, we implemented the Flexible-Budget method (FBM) proposed by [32]. For a fair comparison, we implemented their method using the same NSGA-II parameters used by EMOPaT, including the budget of LL-EA evaluations. The secondary criterion used to compare equally ranked individuals (see [32] for more details) is the Area Under the Curve. Ten independent runs of FBM were run, after which we selected the solutions that among all runs, had obtained the best-performing configurations after 1 K, 10 K, 100 K evaluations (similar to our setting) and 5.5 K and 55 K evaluations (for a comparison with our inferred parameter sets). Then, we performed 100 runs for each configuration and compared the results to the ones reported in Table 4 (for intermediate values we used the ones called "From Pareto") allowing the same computational budget. Table 5 shows the parameters found by FBM and Table 6 the comparison between the performance of the two methods. Except for PSO on the Sphere function, for which the results provided by EMOPaT are always better (Wilcoxon signed-rank test, $p < 0.01$), the two tuning methods always obtain equivalent results with budgets of 1 K, 10 K and 100 K evaluations. With the two intermediate budgets, FBM is better three times, EMOPaT is better twice and once they are equivalent; therefore no significant difference between the two methods can be observed.

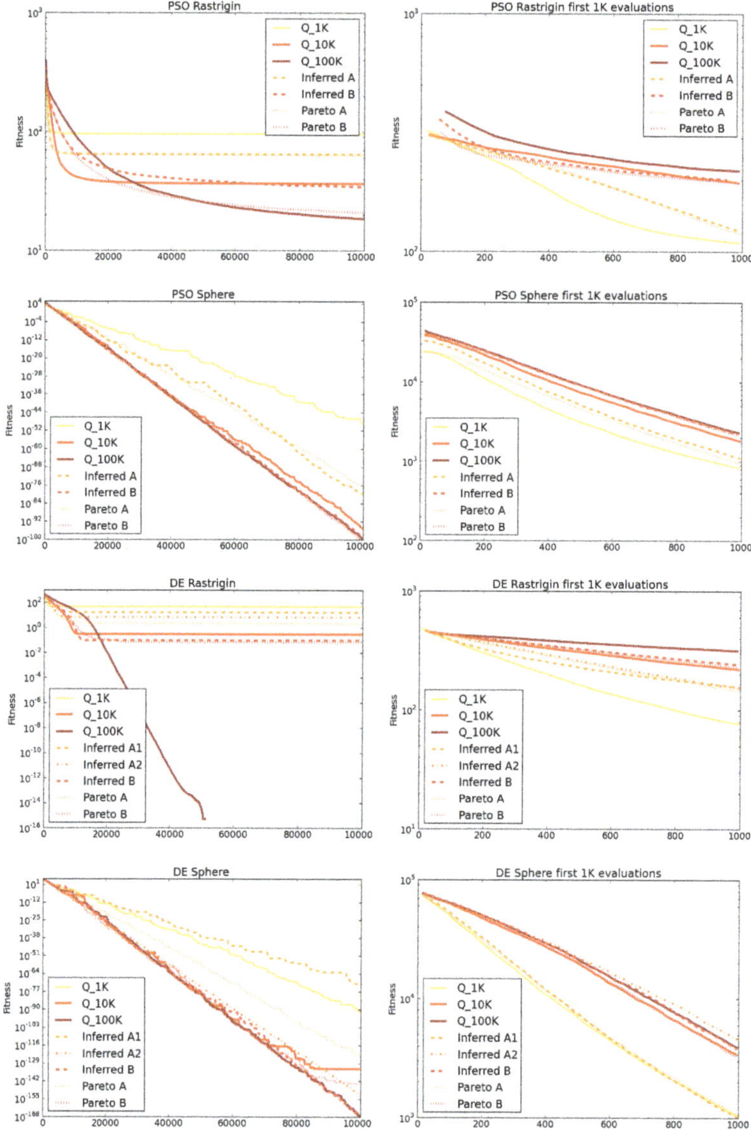

Figure 7. Average fitness versus number of fitness evaluations for configurations generated for PSO (first and second rows) and DE (third and fourth) for the 30-dimensional Rastrigin (above) and Sphere (below) functions. The plots on the right magnify the first 1000 evaluations to better compare the performance of the "fast" versions.

The results obtained by EMOPaT are also consistent with previous experimental and theoretical findings, such as the ones reported in [39], which proved that premature convergence can be avoided if:

$$F > \sqrt{\frac{1 - \frac{CR}{2}}{PopSize}} \tag{7}$$

Of the fourteen DE instances in Table 4, Sphere Q_{1K} is the only one that does not respect this condition; this may be an explanation for the fact that this DE version is fast, but generally unable to reach convergence optimizing such a simple unimodal function.

Table 5. DE and PSO configurations obtained by the Flexible-Budget Method [32].

		Differential Evolution				
	Budget	**PopSize**	**CR**	**F**	**Mutation**	**Crossover**
Rastrigin	1 K	5	0.375	0.375	*random*	*exponential*
	5.5 K	9	0.182	0.182	*random*	*exponential*
	10 K	16	0.145	0.145	*random*	*exponential*
	55 K	24	0.146	0.146	*random*	*exponential*
	100 K	75	0.746	0.746	*random*	*exponential*
Sphere	1 K	16	0.299	0.554	*target-to-best*	*binomial*
	5.5 K	7	0.069	0.751	*target-to-best*	*exponential*
	10 K	7	0.056	0.749	*target-to-best*	*exponential*
	55 K	7	0.057	0.749	*target-to-best*	*exponential*
	100 K	7	0.057	0.749	*target-to-best*	*exponential*
		Particle Swarm Optimization				
	Budget	**PopSize**	**w**	c_1	c_2	**Topology**
Rastrigin	1 K	11	0.664	1.518	0.638	*global*
	5.5 K	88	0.662	1.062	0.624	*global*
	10 K	119	0.654	1.672	0.607	*global*
	55 K	230	0.650	2.412	0.643	*global*
	100 K	230	0.650	2.412	0.643	*global*
Sphere	1 K	17	0.837	0.952	0.488	*global*
	5.5 K	32	0.897	0.539	0.545	*global*
	10 K	32	0.897	0.539	0.545	*global*
	55 K	34	0.380	1.135	2.673	*global*
	100 K	34	0.380	1.135	2.673	*global*

Table 6. Comparison between the performance of the configurations found by EMOPaT and the Flexible-Budget method (FBM).

		Differential Evolution			
	Budget	**EMOPaT**	**FBM**	**p-Value**	**Best Method**
Rastrigin	1 K	$7.68E + 01$	$7.76E + 01$	$1.42E - 01$	
	5.5 K	$4.29E + 00$	$5.06E + 00$	$8.53E - 03$	EMOPaT
	10 K	$3.32E - 02$	$1.04E - 01$	$4.49E - 01$	
	55 K	$5.68E - 14$	$0.00E + 00$	$1.43E - 08$	FBM
	100 K	0.0	0.0	$8.33E - 02$	
Sphere	1 K	$1.11E + 02$	$1.08E + 02$	$9.62E - 03$	EMOPaT
	5.5 K	$6.51E + 01$	$5.32E + 01$	$3.62E - 11$	FBM
	10 K	$4.24E + 01$	$4.15E + 01$	$6.52E - 01$	
	55 K	$2.45E + 01$	$2.53E + 01$	$1.28E - 01$	
	100 K	0.0	0.0	$2.49E - 01$	
		Particle Swarm Optimization			
	Budget	**EMOPaT**	**FBM**	**p-Value**	**Best Method**
Rastrigin	1 K	$8.77E + 02$	$1.04E + 03$	$9.51E - 01$	
	5.5 K	$5.10E - 04$	$5.19E - 05$	$3.23E - 05$	FBM
	10 K	$6.13E - 13$	$5.69E - 13$	$4.98E - 01$	
	55 K	$2.64E - 93$	$4.27E - 93$	$5.58E - 03$	EMOPaT
	100 K	$1.79E + 01$	$2.03E + 01$	$7.05E - 02$	
Sphere	1 K	$7.16E + 02$	$1.40E + 03$	$1.33E - 13$	EMOPaT
	5.5 K	$1.03E - 02$	$8.18E + 00$	$3.90E - 18$	EMOPaT
	10 K	$1.96E - 07$	$1.11E - 01$	$3.90E - 18$	EMOPaT
	55 K	$1.46E - 56$	$2.23E - 22$	$3.86E - 18$	EMOPaT
	100 K	0.0	$5.15E - 44$	$3.88E - 18$	EMOPaT

5.2. Multi-Function Optimization

In this section, we show how EMOPaT behaves when the goal is to obtain configurations that perform well on functions that are not included in the "training set". Following the terminology introduced by [31], we expect to find "generalist" and "specialist" versions of the EAs taken into consideration. Table 7 gives more details about this experiment.

Table 7. Optimization of seven different functions. Experimental settings.

EMOPaT settings
Population Size = 200, 80 Generations, Mutation Rate = 0.125, Crossover Rate = 0.9
Function settings
10-dimensional Sphere, Rotated Cigar, Rosenbrock, Rotated Ackley,
Rastrigin, Composition Function 1 (CF1), Composition Function 3 (CF3)
evaluation criterion = best fitness in 20000 evaluations averaged over $N = 15$ repetitions.

We used EMOPaT to optimize all the seven functions together (repeating the test 10 times) and then we merged all results into a single collection of configurations. From this collection, we selected the best-performing configuration for each of the seven objective functions.

The next step was to select the "generalist" solutions. We consider a "generalist" solution to be a parameter set that does not perform badly on any of the objectives taken into consideration, i.e., it is never in the worst θ percent of the population, when ordered by any objective. Obviously, the higher θ, the lower the number of generalists. We decided to set the value of θ such that seven generalists would be selected, to match the specialists' number.

Table 8 shows the generalists' and specialists' parameters obtained by merging the results of ten independent runs of EMOPaT. An interesting outcome worth highlighting is that, similar to the previous experiment, some of the generalists are not obtained by simply "interpolating" other results but they contain some traits that are not featured by any specialist. For instance, DE G_0 has a smaller population than any specialist, PSO G_3's inertia value is higher than that of all specialists.

A more standard way to infer a "generalist" configuration is to take the one with the best overall results. To do so, we consider the results of all the solutions found by EMOPaT and normalize them so that each fitness has average = 0 and standard deviation = 1; then, we select the configuration that minimizes the sum of the normalized fitnesses. In Table 8, these configurations are reported as "average".

We also performed 10 meta-optimizations using SEPaT and *irace*, with the same budget allowed for EMOPaT. The parameters of SEPaT are the ones presented in Table A1, while for *irace* we used the parameters suggested by the authors. For each optimization method, the ten solutions obtained were compared using the tournament method described in Section 4 to find the best configuration, which is also reported in Table 8.

To test the parameter sets obtained, we selected seven functions from the CEC 2013 benchmark that were not used during training (namely Elliptic, Rotated Discus, Rotated Weierstrass, Griewank, Rotated Katsuura, CF5 and CF7). Table 9 shows, for each function, which configuration(s) obtained the best results. To determine the best function, we performed the Wilcoxon signed-rank test ($p < 0.01$) on all configurations pairwise. A configuration is considered to be the best if no other configuration performs significantly better on that function. The table shows that, in some cases, generalists were actually able to obtain better results on previously unseen functions than specialists.

Since the definition of "generalist EA" implies the ability not to perform badly on any function, we also analyzed the same data from another viewpoint. Each cell (i, j) in Table 10 shows the number of test functions for which the optimizer which row i refers to performs statistically worse than the one referred to by column j (Wilcoxon signed-rank test, $p < 0.01$). The last column reports the sum of each line and can be considered an indicator of the generalization ability of the optimizer with respect to the others over the test functions.

It can be observed that some of the generalists performed very well. The best optimizers for DE were the configurations obtained by SEPaT along with two generalists, G_4 and G_5. The first two configurations are very similar to each other (same mutation and crossover, $CR \simeq 0.15$ and $F \simeq 0.5$), as shown by the presence of statistically significant differences between them only on one function out of seven. No specialist features a similar parameter set. Regarding PSO, two of the specialists (S_{cigar} and $S_{rosenbrock}$) obtained very good results, as well as three of the generalists (G_0, G_1, and G_3). It is important to notice that most generalists evolved by EMOPaT outperform the solutions found by the other single-objective tuners used as reference, as well as the one obtained by computing a normalized average of all solutions evolved by EMOPaT ("average" in Table 8). This last configuration (which is the same as S_{ackley} for PSO) was the best optimizer for three functions and the worst one (not reported) for two (Elliptic, Katsuura). This suggests that this is not the correct way of finding a configuration able to perform well on different functions.

In conclusion, we can say that EMOPaT, in a single optimization process, is able to find, at the same time, algorithm configurations that work well on a single function of interest and others that are able to generalize over different unseen functions, while single-objective tuners need separate processes with a consequent increase of the time spent to perform this operation.

Table 8. The seven DE and PSO best-performing configurations generated by EMOPaT for each "training" function (denoted by S, for specialist, followed by the name of the function); the seven configurations that never achieved bad results in any of them (denoted by G, for generalist); the parameter sets found by *irace* and by SEPaT; and the single generalist configuration obtained by normalizing fitness values (see text).

Differential Evolution					
Configuration Name	**PopSize**	**CR**	**F**	**Mutation**	**Crossover**
S_{sphere}	12	0.181	0.718	*target-to-best*	*exponential*
S_{cigar}	57	0.906	0.703	*target-to-best*	*exponential*
$S_{rosenbrock}$	47	0.989	0.761	*random*	*exponential*
S_{ackley}	271	0.170	0.216	*target-to-best*	*exponential*
$S_{rastrigin}$	24	0.024	1.158	*random*	*exponential*
S_{CF1}	24	0.057	1.789	*best*	*exponential*
S_{CF3}	98	0.868	0.087	*random*	*binomial*
G_0	10	0.607	0.886	*target-to-best*	*exponential*
G_1	70	0.612	0.480	*best*	*exponential*
G_2	13	0.235	0.444	*target-to-best*	*exponential*
G_3	23	0.413	0.860	*target-to-best*	*exponential*
G_4	32	0.147	0.491	*target-to-best*	*exponential*
G_5	24	0.776	0.716	*target-to-best*	*exponential*
G_6	19	0.058	0.837	*best*	*binomial*
irace	53	0.796	0.508	*best*	*exponential*
SEPaT	17	0.160	0.499	*target-to-best*	*exponential*
average	40	0.563	0.988	*target-to-best*	*binomial*

Particle Swarm Optimization					
Configuration Name	**PopSize**	**w**	c_1	c_2	**Topology**
S_{sphere}	34	0.768	1.756	0.474	*global*
S_{cigar}	41	0.585	1.338	1.646	*ring*
$S_{rosenbrock}$	55	-0.465	-0.060	1.930	*ring*
S_{ackley}	251	0.714	1.082	0.271	*global*
$S_{rastrigin}$	20	-0.131	-0.050	3.787	*global*
S_{CF1}	161	-0.158	-0.112	2.467	*ring*
S_{CF3}	269	-0.172	1.235	1.945	*global*
G_0	43	0.648	1.241	1.633	*ring*
G_1	44	0.639	2.114	1.478	*ring*
G_2	68	0.402	1.109	2.184	*ring*
G_3	37	0.883	0.548	0.649	*ring*
G_4	26	-0.073	0.295	3.032	*ring*
G_5	33	0.583	2.040	1.677	*ring*
G_6	46	0.593	1.944	1.637	*ring*
irace	19	0.805	0.962	0.914	*ring*
SEPaT	22	0.732	1.358	1.153	*ring*
average	251	0.714	1.082	0.271	*global*

Table 9. Best-performing DE and PSO configurations on the seven test functions.

Function	DE	PSO
Elliptic	S_{sphere}	*irace*, SEPaT
Discus	S_{cigar}	average, S_{ackley}
Weierstrass	S_{cigar}, *irace*	S_{sphere}, $S_{rosenbrock}$
Griewank	$S_{rastrigin}$, SEPaT, G_4	G_0, G_1, G_2, G_5, G_6, S_{cigar}
Katsuura	G_2, SEPaT, G_4	S_{CF1}, $S_{rosenbrock}$
CF5	S_{cigar}, SEPaT	S_{ackley}, average
CF7	G_2, SEPaT, G_4	S_{ackley}, $S_{rosenbrock}$, average

Table 10. Number of test functions for which the optimizer associated with the row is statistically worse than the one associated with the column. The last column reports the sum of the values in that row, measuring the optimizer performance (the lower, the better). The three best configurations are highlighted in bold.

	S_{sphere}	S_{cigar}	$S_{rosenbrock}$	S_{ackley}	$S_{rastrigin}$	S_{CF1}	S_{CF3}	G_0	G_1	G_2	G_3	G_4	G_5	G_6	*irace*	SEPaT	average	sum
Differential Evolution																		
S_{sphere}	0	4	4	4	3	0	4	4	4	5	3	6	4	2	4	6	2	59
S_{cigar}	2	0	0	1	1	1	1	3	1	4	2	3	2	2	2	4	1	30
$S_{rosenbrock}$	3	6	0	2	4	2	2	3	4	4	3	4	5	3	4	4	2	55
S_{ackley}	3	4	4	0	3	2	2	4	4	5	3	5	4	3	4	5	4	59
$S_{rastrigin}$	1	3	2	2	0	0	2	3	3	5	3	6	3	3	3	5	1	45
S_{CF1}	4	5	3	4	7	0	4	4	7	7	5	7	7	5	6	7	3	85
S_{CF3}	2	5	3	3	3	2	0	4	3	4	3	4	3	3	4	4	2	52
G_0	2	4	4	1	2	0	3	0	4	3	1	4	6	2	5	4	2	47
G_1	2	4	2	0	1	0	2	1	0	4	2	4	4	2	3	4	2	37
G_2	2	3	2	1	1	0	3	2	3	0	0	3	4	1	3	2	1	31
G_3	2	4	3	2	2	0	4	3	4	3	0	6	5	1	4	6	3	52
G_4	1	3	3	1	0	0	2	3	1	1	1	0	3	1	3	1	1	**25**
G_5	1	3	1	0	1	0	1	0	0	3	1	3	0	1	2	3	0	**20**
G_6	1	4	4	2	2	0	4	2	4	2	0	5	4	0	4	4	2	44
irace	1	3	1	1	2	1	0	1	2	3	1	3	3	1	0	3	1	27
SEPaT	1	2	2	1	0	0	2	2	1	0	0	1	3	1	3	0	1	**20**
average	2	5	3	3	3	1	4	3	4	4	3	6	5	2	5	5	0	58
Particle Swarm Optimization																		
S_{sphere}	0	3	4	3	2	2	1	3	3	3	3	2	3	3	2	3	3	43
S_{cigar}	4	0	4	4	0	4	3	0	2	0	1	0	1	2	1	1	4	**31**
$S_{rosenbrock}$	1	2	0	2	0	0	2	2	2	2	3	1	2	2	2	2	2	**27**
S_{ackley}	3	3	3	0	3	2	2	3	3	3	3	3	3	3	3	3	0	43
$S_{rastrigin}$	5	6	7	4	0	4	5	6	6	6	6	5	6	6	6	6	4	88
S_{CF1}	2	2	3	5	1	0	3	2	2	2	2	2	2	2	2	2	5	39
S_{CF3}	4	3	4	4	2	2	0	3	2	2	3	3	3	3	3	3	4	48
G_0	4	1	4	4	0	3	2	0	2	0	1	0	1	2	1	1	4	**30**
G_1	1	2	3	4	1	3	1	2	0	1	3	2	1	1	2	1	3	**31**
G_2	4	1	4	4	0	4	3	1	1	0	2	1	2	2	1	1	4	35
G_3	2	2	3	4	0	3	2	2	1	1	0	1	2	2	1	1	4	**31**
G_4	4	4	4	4	0	5	4	4	3	3	4	0	4	3	4	4	4	58
G_5	3	4	3	4	1	4	2	1	1	2	3	1	0	1	3	3	4	40
G_6	2	2	3	4	1	3	1	2	0	1	2	2	1	0	2	2	4	32
irace	4	2	4	4	1	4	3	2	3	3	3	2	3	3	0	1	4	46
SEPaT	4	2	4	4	1	4	3	2	2	2	2	1	2	3	0	0	4	40
average	3	3	3	0	3	2	2	3	3	3	3	3	3	3	3	3	0	43

6. Summary and Future Work

In this paper we presented some examples of the kind of information and insights into stochastic optimization algorithms that can be offered by a multi-objective meta-optimization environment. To do so, we used EMOPaT, a simple and generic multi-objective evolutionary optimization framework for tuning the parameters of an EA. EMOPaT was tested on the optimization of DE and PSO in different scenarios, showing that it is able to highlight how the parameters affect the performance of an EA in different situations, allowing one to draw generalizable results when considering different constraints applied to the optimization of the same function. Successively, we tested it on different functions and proved it not only allows one to find good configurations for the training function(s), but also to derive from those results new parameter sets that perform well on unseen problems.

We think that EMOPaT can be helpful in many applications and provide useful hints about the behavior of any metaheuristic. In [40] we showed that EMOPaT can be effective in real-world situations, by using it to tune a DE-based object recognition algorithm. In general, a basic application of EMOPaT can be summarized in the following steps, as described also in the code we made available online:

1. Select a proper set of problem-related fitness cases.
2. Select the optimization method(s) whose parameters one wants to tune.
3. Select the objectives to optimize (convergence speed, solution quality, robustness, etc.).
4. Run EMOPaT and save the resulting parameter set.

Below we report some interesting directions towards which this approach can be further expanded:

- At present, the analysis of the results is essentially a "manual" process. Which is the best way to automatically extract, generalize and infer parameters?
- EMOPaT belongs to the class of offline parameter tuning algorithms, in which the values of the parameters are set before starting the optimization process and do not change during its execution. Could it also be used to tune the parameters of a population of EA's online, adapting parameter values as optimization proceeds?
- In our work, we proved that EMOPaT can also be used to generalize results on a single function (see Section 5.1). Can this idea be extended to different functions? If we obtain a Pareto Front by optimizing two functions, can we extract parameters for a function that lies "between" these two, according to some metric that takes into consideration some of their properties?
- Is it possible to group or cluster functions based on the best-performing parameters found by EMOPaT?

Author Contributions: Conceptualization, R.U. and S.C.; Investigation, R.U., L.S. and S.C.; Software, R.U.; Supervision, S.C.

Funding: This research received no external funding.

Ethical Approval: This article does not contain any studies with human participants performed by any of the authors.

Conflicts of Interest: The authors declare no conflict of interest.

Appendix A

In this appendix, we first show that EMOPaT can be considered a generalization of its single-objective version SEPaT. This implies that for single-objective optimization problems, we can extend the same conclusions drawn for SEPaT in [5,41] to EMOPaT. Then, to assess its general soundness, we demonstrate EMOPaT's ability to give insights about the algorithm parameters and on their influence on the optimization process by showing that EMOPaT can correctly deal with some peculiar situations, such as the presence of useless or bad parameters.

Appendix A.1. Comparison with SEPaT

The equivalence between SEPaT and EMOPaT in the single-objective case has been tested on seven functions (see Table A1) from the CEC 2013 benchmark [38], with the only difference that the function minima were set to 0.

Table A1. Comparison between EMOPaT and SEPaT. Experimental settings.

EMOPaT settings*   Tuner EA = NSGA-II, Population Size = 200, 80 Generations,   Mutation Rate = 0.125, Crossover Rate = 0.9
SEPaT settings   Tuner EA = DE, Population Size = 200, 80 Generations   CR = 0.91, F = 0.52, Mutation = *target-to-best*, Crossover = *Exponential*
Function settings   10-dimensional Sphere, Rotated Cigar, Rotated Rosenbrock, Rotated Ackley,   Rastrigin, Composition Function 1 (CF1), Composition Function 3 (CF3)   evaluation criterion = best fitness in 20,000 evaluations averaged over $N = 15$ repetitions.

First, we performed tuning as a single run of EMOPaT considering these functions as seven different objectives (optimizing all the functions together), and then by running seven times SEPaT, once for each function. More details about these experiments are summarized in Table A1.

We checked whether the best solutions for each objective that EMOPaT evolved in a single run (also called "top solutions" or "top configurations" in the following), were actually indistinguishable from those obtained by SEPaT when applied to the same objective. To do so, we ran ten independent experiments with both SEPaT (once for each function) and EMOPaT. The best EA configuration for each function found in each run was then tested 100 times on the optimization of the corresponding function. We computed the median for each set of 100 tests and, based on it, selected the overall best configuration for each function.

Table A2 compares the best PSO and DE configurations obtained by SEPaT in ten independent runs to the best configurations obtained, for each corresponding function, in ten independent runs of EMOPaT; the parameters obtained by the two methods are significantly similar. For instance, the nominal parameters chosen for both DE and PSO are almost always the same except for the PSO topology for Composition Function 3. This is the only case in which the parameters chosen by the two methods are clearly different (one population is three times as large as the other, c_1 is four times larger and the topology is different): nevertheless, the results obtained by the two configurations are virtually equivalent (see Table A3), so the two settings correspond to two equivalent minima of the meta-fitness landscape.

Table A3 shows the median fitness obtained on each function by the best-performing EA configurations found by the tuners and by a standard configuration, and the p-values of Wilcoxon's signed-rank test under the Null Hypothesis "There are no differences between two configurations' performance" comparing EMOPaT's best configuration to SEPaT's best and to a standard configuration. While, in general, EMOPaT's configurations perform better than standard parameters (last column), there is no statistical evidence that the best performance of the configurations found by the two methods differ, except for two cases (Rotated Cigar and Rotated Ackley using DE) for which EMOPaT performs slightly better than SEPaT. These results show that EMOPaT can be thought as being generally equivalent to SEPaT in finding the minima of single-objective problems. However, as shown in the paper, one can extract even more information from EMOPaT's results, thanks to its multi-objective nature.

Table A2. Best-performing parameters obtained over 10 runs of EMOPaT and SEPaT, and standard settings for PSO ([42]) and DE ([11]).

			Differential Evolution			
Function	Method	PopSize	CR	F	Mutation	Crossover
Sphere	SEPaT	20	0.506	0.520	*target-to-best*	*exponential*
	EMOPaT	12	0.181	0.718	*target-to-best*	*exponential*
R. Cigar	SEPaT	60	0.955	0.660	*target-to-best*	*binomial*
	EMOPaT	38	0.916	0.699	*target-to-best*	*binomial*
R. Rosenbrock	SEPaT	39	0.993	0.745	*random*	*exponential*
	EMOPaT	47	0.989	0.761	*random*	*exponential*
R. Ackley	SEPaT	85	0.327	0.0	*random*	*exponential*
	EMOPaT	248	0.960	0.0	*random*	*exponential*
Rastrigin	SEPaT	36	0.014	0.359	*random*	*exponential*
	EMOPaT	25	0.049	1.065	*random*	*exponential*
CF 1	SEPaT	18	0.0	1.777	*best*	*exponential*
	EMOPaT	33	0.045	1.070	*best*	*exponential*
CF 3	SEPaT	89	0.794	0.070	*random*	*binomial*
	EMOPaT	98	0.868	0.088	*random*	*binomial*
-	Standard	30	0.9	0.5	*random*	*exponential*

			Particle Swarm Optimization			
Function	Method	PopSize	w	c_1	c_2	Topology
Sphere	SEPaT	88	0.529	1.574	1.057	*global*
	EMOPaT	25	0.774	1.989	0.591	*global*
R. Cigar	SEPaT	67	0.713	0.531	1.130	*ring*
	EMOPaT	41	0.757	1.159	1.097	*ring*
R. Rosenbrock	SEPaT	104	0.597	1.032	1.064	*ring*
	EMOPaT	87	−0.451	−0.092	1.987	*ring*
R. Ackley	SEPaT	113	0.381	0.210	1.722	*ring*
	EMOPaT	115	0.303	−0.006	2.467	*ring*
Rastrigin	SEPaT	13	−0.236	0.090	3.291	*global*
	EMOPaT	7	−0.260	0.021	3.314	*global*
CF 1	SEPaT	92	−0.147	−0.462	2.892	*ring*
	EMOPaT	61	−0.163	−0.376	3.104	*ring*
CF 3	SEPaT	61	0.852	0.347	0.989	*ring*
	EMOPaT	217	0.728	1.217	0.565	*global*
-	Standard	30	0.721	1.193	1.193	*ring*

Table A3. Median fitness over 100 independent runs of the best solutions found by EMOPaT, by SEPaT, and by a standard configuration of the optimization algorithm.

EA	Function	EMOPaT	SEPaT	Standard	vs. SEPaT	vs. Standard
			Fitness		**p-Value**	
DE	Sphere	0.00	0.00	$7.43E-26$	1.00	$<1E-20$
	R. Cigar	$7.61E-02$	$7.64E-04$	$1.76E+01$	$4.89E-03$	$5.49E-08$
	R. Rosenbrock	$3.42E-02$	$2.76E-03$	$9.81E+00$	0.41	$<1E-20$
	R. Ackley	$2.04E+01$	$2.05E+01$	$2.05E+01$	$1.21E-03$	$9.34E-07$
	Rastrigin	0.00	0.00	$2.17E-08$	1.00	$<1E-20$
	CF 1	$2.04E+02$	$2.05E+02$	$4.00E+02$	0.06	$<1E-20$
	CF 3	$6.09E+02$	$6.14E+02$	$1.46E+03$	0.67	$<1E-20$
PSO	Sphere	0.00	0.00	$7.57E-24$	1.00	$<1E-20$
	R. Cigar	$1.33E+06$	$2.43E+06$	$1.84E+06$	0.05	0.03
	R. Rosenbrock	$9.44E-01$	$1.01E+00$	$9.81E+00$	0.87	$1.04E-17$
	R. Ackley	$2.04E+01$	$2.04E+01$	$2.04E+01$	0.16	0.67
	Rastrigin	$4.75E-06$	$1.02E-04$	$1.09E+01$	0.57	$5.88E-08$
	CF 1	$2.01E+02$	$2.03E+02$	$4.00E+02$	0.99	$<1E-20$
	CF 3	$9.62E+02$	$9.85E+02$	$1.16E+03$	0.27	$2.7E-04$

Appendix A.2. Empirical Validation

We have artificially created four test cases characterized by:

1. A useless numerical parameter, i.e., with no effects at all on the algorithm;
2. A harmful numerical parameter, i.e., the higher its value, the worse the fitness;
3. A harmful nominal parameter choice that constantly produces bad fitness values when made;
4. Two totally equivalent choices of a nominal parameter.

A similar approach has been proposed by [43], showing the ability of *irace*, ParamILS and REVAC to recognize an operator which was detrimental for the fitness. The results of these tests increase the confidence in the actual ability of EMOPaT to recognize the usefulness or, more in general, the role of a parameter of an EA. We limited our tests to optimizing PSO on the Sphere and Rastrigin functions (see Table A4). In these tests, we modified the original encoding of PSO configurations (Figure 3) as shown in Figure A1.

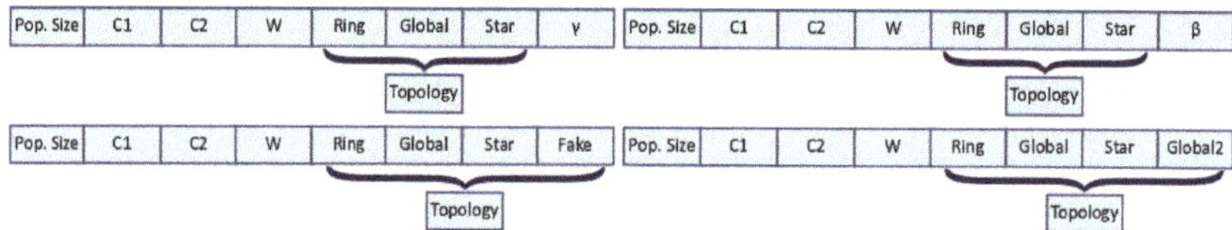

Figure A1. Encoding of PSO configurations in the four cases presented in Appendix A.2. From top left clockwise: useless parameter, harmful numerical parameter, equivalent and harmful topology.

Table A4. Empirical validation of EMOPaT. Experimental settings.

EMOPaT settings
Population Size = 64, 100 Generations, Mutation Rate = 0.125, Crossover Rate = 0.9

Function settings
30-dimensional Sphere and Rastrigin
evaluation criterion = best fitness in 20000 evaluations averaged over $N = 15$ repetitions

Appendix A.2.1. Useless Parameter

In this experiment, we extended the PSO configuration encoding by adding a parameter γ that does not appear in the algorithm and therefore has no effects on it. Our goal was to analyze how EMOPaT dealt with such a parameter with respect to the actually effective ones. Table A5 shows mean and standard deviation of the (normalized) numerical parameters in all NSGA-II individuals at the end of ten independent runs. As can be observed, the useless parameter γ has a mean value close to 0.5 and its variance is 0.078, which is very close to $\frac{1}{12}$, expected for a uniform distribution in $[0, 1]$: this does not happen with the other parameters. Figure A2 plots the values of the Sphere function against the values of PSO parameters (after recovering their actual value). While the values of the real parameters show a clear trend, the values of γ are scattered uniformly all over the graph. As well, the correlation of γ with the other numerical parameters is very low (last row of Table A5). This suggests that a useless parameter can be easily identified by a (quasi-)uniform distribution of its values.

Table A5. Mean and variance values for PSO's numerical parameters and correlation with a useless one (γ). Parameter values are normalized between 0 and 1.

Parameter	Population Size	w	C_1	C_2	γ
Mean	0.159	0.555	0.586	0.332	0.441
Variance	0.0313	0.0109	0.0120	0.0103	0.0780
Correlation with γ	-0.0777	-0.179	-0.159	0.149	-

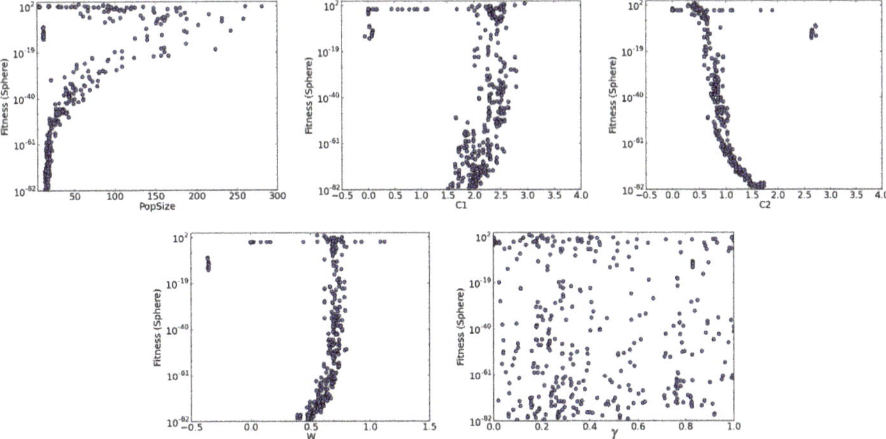

Figure A2. Values of fitness (Sphere function) versus PSO parameters at the end of the tuning procedure. The last graph refers to the useless parameter γ which, unlike the others, spans across all possible values with no correlation with fitness.

Appendix A.2.2. Harmful Numerical Parameter

In this experiment, we added to the representation of each PSO configuration a parameter $\beta \in [0, 1]$ whose only effect is to worsen the actual fitness f proportionally to its value as follows:

$$\hat{f} = (f + \beta) \cdot (1 + \beta) \tag{A1}$$

Parameter β was constantly assigned values close to 0 (mean 7×10^{-4}, variance 7×10^{-6}) by EMOPaT. Figure A3 plots values of β versus number of generations, averaged over ten EMOPaT runs. β starts from an average of 0.5 (due to random initialization) but, after a few iterations, its value quickly reaches 0.

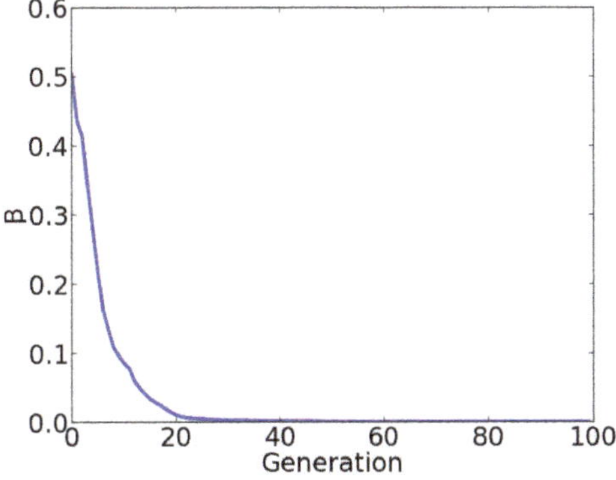

Figure A3. Evolution of the "bad parameter" β, averaged over all individuals in ten independent runs of EMOPaT, versus generation number.

Appendix A.2.3. Harmful Nominal Parameter Setting

In this experiment, we added a "fake" fourth topology to PSO configurations. When it is selected, PSO just returns a bad fitness value. We wanted to verify whether that choice would be always discarded and which values the corresponding gene would take. Figure A4 shows that the fake topology is actually discarded and, after only two generations, is never selected anymore. Moreover, the values of the corresponding gene are always lower than all others; in particular, they are lower than the ones representing the *star* topology, which is also never selected despite being a valid choice.

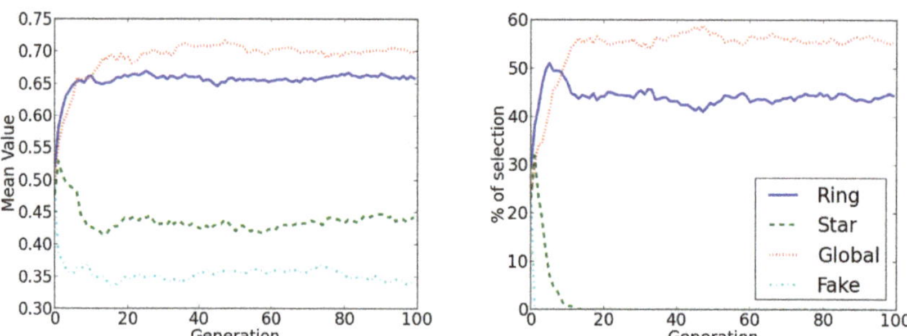

Figure A4. Average values and selection percentages of the genes representing the four topologies versus number of EMOPaT generations. Results averaged over 64 individuals in 10 runs.

Appendix A.2.4. Equivalent Settings

In the last experiment of this section, we added to the basic representation of the PSO configuration a fourth topology that when selected, acts exactly as the *global* topology. Our goal was to see whether EMOPaT would allow one to understand that the two topologies were in fact the same one. Figure A5 shows the results in the same format as Figure A4. There is no clear correlation between the two "global" versions, but it can be observed that at the end of the evolution, the sum of their selection percentages has converged to the value reached by *global* in the previous experiment. This means that splitting this choice into two distinct values did not affect EMOPaT's performance. Nevertheless, these results were reached more slowly, showing that it takes time for EMOPaT to reach the correct values of a nominal parameter when many choices are available.

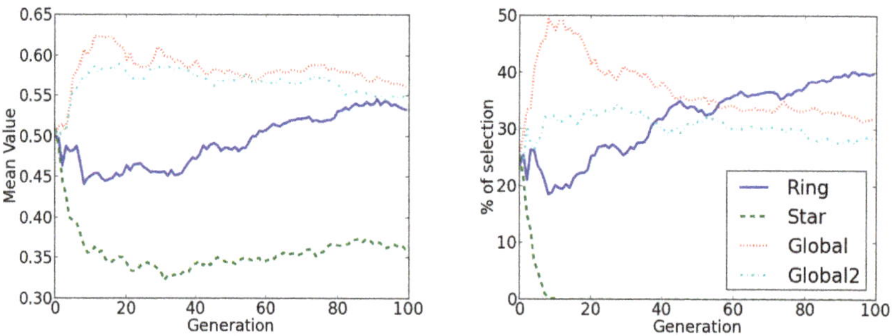

Figure A5. Average values of the genes representing the four topologies (including the replicated one) and selection percentages. The *x* axis reports the number of EMOPaT generations.

References

1. Eiben, A.E.; Smith, J.E. *Introduction to Evolutionary Computing*; Springer: Berlin, Germany, 2003.
2. Pitzer, E.; Affenzeller, M. A Comprehensive Survey on Fitness Landscape Analysis. In *Recent Advances in Intelligent Engineering Systems*; Studies in Computational Intelligence; Fodor, J., Klempous, R., Suárez Araujo, C., Eds.; Springer: Berlin/Heidelberg, Germany, 2012; Volume 378, pp. 161–191.

3. Smith-Miles, K.; Tan, T. Measuring algorithm footprints in instance space. In Proceedings of the IEEE Congress on Evolutionary Computation (CEC), Brisbane, QLD, Australia, 10–15 June 2012; pp. 1–8.

4. Mercer, R.; Sampson, J. Adaptive search using a reproductive metaplan. *Kybernetes* **1978**, *7*, 215–228. [CrossRef]

5. Ugolotti, R.; Nashed, Y.S.G.; Mesejo, P.; Cagnoni, S. Algorithm Configuration using GPU-based Metaheuristics. In Proceedings of the Genetic and Evolutionary Computation Conference (GECCO), Amsterdam, The Netherlands, 6–10 July 2013; pp. 221–222.

6. Storn, R.; Price, K. *Differential Evolution—A Simple and Efficient Adaptive Scheme for Global Optimization over Continuous Spaces*; Technical Report; International Computer Science Institute: Berkeley, CA, USA, 1995.

7. Kennedy, J.; Eberhart, R. Particle Swarm Optimization. In Proceedings of the IEEE International Conference on Neural Networks, Perth, Australia, 27 November–1 December 1995; Volume 4, pp. 1942–1948.

8. Ugolotti, R.; Cagnoni, S. Analysis of Evolutionary Algorithms using Multi-Objective Parameter Tuning. In Proceedings of the Genetic and Evolutionary Computation Conference (GECCO), Vancouver, BC, Canada, 12–16 July 2014; pp. 1343–1350.

9. Deb, K.; Pratap, A.; Agarwal, S.; Meyarivan, T. A fast and elitist multiobjective genetic algorithm: NSGA-II. *IEEE Trans. Evolut. Comput.* **2002**, *6*, 182–197. [CrossRef]

10. Deb, K.; Srinivasan, A. Innovization: Innovating design principles through optimization. In Proceedings of the Genetic and Evolutionary Computation Conference (GECCO), Seattle, WA, USA, 8–12 July 2006; pp. 1629–1636.

11. Das, S.; Suganthan, P. Differential Evolution: A Survey of the State-of-the-Art. *IEEE Trans. Evolut. Comput.* **2011**, *15*, 4–31. [CrossRef]

12. Montero, E.; Riff, M.C.; Rojas-Morales, N. Tuners review: How crucial are set-up values to find effective parameter values? *Eng. Appl. Artif. Intell.* **2018**, *76*, 108–118. [CrossRef]

13. Sipper, M.; Fu, W.; Ahuja, K.; Moore, J.H. Investigating the parameter space of evolutionary algorithms. *BioData Min.* **2018**, *11*, 2. [CrossRef] [PubMed]

14. Karafotias, G.; Hoogendoorn, M.; Eiben, Á.E. Parameter control in evolutionary algorithms: Trends and challenges. *IEEE Trans. Evolut. Comput.* **2015**, *19*, 167–187. [CrossRef]

15. Nannen, V.; Eiben, A.E. Relevance Estimation and Value Calibration of Evolutionary Algorithm Parameters. In Proceedings of the International Joint Conference on Artifical Intelligence (IJCAI), Hyderabad, India, 6–12 January 2007; pp. 975–980.

16. Larrañaga, P.; Lozano, J.A. *Estimation of Distribution Algorithms: A New Tool for Evolutionary Computation*; Kluwer Academic Publishers: Norwell, MA, USA, 2001.

17. Smit, S.K.; Eiben, A.E. Beating the 'world champion' evolutionary algorithm via REVAC tuning. In Proceedings of the IEEE Congress on Evolutionary Computation (CEC), Barcelona, Spain, 18–23 July 2010; pp. 1–8.

18. Suganthan, P.N.; Hansen, N.; Liang, J.J.; Deb, K.; Chen, Y.P.; Auger, A.; Tiwari, S. *Problem Definitions and Evaluation Criteria for the CEC 2005 Special Session on Real-Parameter Optimization*; Technical Report, KanGAL Report 2005005; Nanyang Technological University: Singapore, 2005.

19. Meissner, M.; Schmuker, M.; Schneider, G. Optimized Particle Swarm Optimization (OPSO) and its application to artificial neural network training. *BMC Bioinform.* **2006**, *7*, 125. [CrossRef] [PubMed]

20. Pedersen, M.E.H. Tuning and Simplifying Heuristical Optimization. Master's Thesis, University of Southampton, Southampton, NY, USA, 2010.

21. Hutter, F.; Hoos, H.H.; Leyton-Brown, K.; Stützle, T. ParamILS: An Automatic Algorithm Configuration Framework. *J. Artif. Intell. Res.* **2009**, *36*, 267–306. [CrossRef]

22. Luke, S.; Talukder, A.K.A. Is the meta-EA a Viable Optimization Method? In Proceedings of the Genetic and Evolutionary Computation Conference (GECCO), Amsterdam, The Netherlands, 6–10 July 2013; pp. 1533–1540.

23. Fister, D.; Fister, I.; Jagrič, T.; Brest, J. A novel self-adaptive differential evolution for feature selection using threshold mechanism. In Proceedings of the 2018 IEEE Symposium Series on Computational Intelligence (SSCI), Bengaluru, India, 18–21 November 2018; pp. 17–24.

24. Bartz-Beielstein, T.; Lasarczyk, C.; Preuss, M. Sequential parameter optimization. In Proceedings of the IEEE Congress on Evolutionary Computation (CEC), Edinburgh, UK, 2–4 September 2005; pp. 773–780.

25. López-Ibáñez, M.; Dubois-Lacoste, J.; Cáceres, L.P.; Birattari, M.; Stützle, T. The irace package: Iterated racing for automatic algorithm configuration. *Oper. Res. Perspect.* **2016**, *3*, 43–58. [CrossRef]

26. Pereira, I.; Madureira, A. Racing based approach for Metaheuristics parameter tuning. In Proceedings of the 10th Iberian Conference on Information Systems and Technologies (CISTI), Aveiro, Portugal, 17–20 June 2015; pp. 1–6.

27. Sinha, A.; Malo, P.; Xu, P.; Deb, K. A Bilevel Optimization Approach to Automated Parameter Tuning. In Proceedings of the 2014 Annual Conference on Genetic and Evolutionary Computation (GECCO), Vancouver, BC, Canada, 12–16 July 2014; pp. 847–854.

28. Andersson, M.; Bandaru, S.; Ng, A.; Syberfeldt, A. Parameter Tuning of MOEAs Using a Bilevel Optimization Approach. *Evolutionary Multi-Criterion Optimization*; Springer: Berlin/Heidelberg, Germany, 2015; pp. 233–247.

29. Dréo, J. Using Performance Fronts for Parameter Setting of Stochastic Metaheuristics. In Proceedings of the Conference on Genetic and Evolutionary Computation Conference (GECCO): Late Breaking Papers, Montreal, QC, Canada, 8–12 July 2009; pp. 2197–2200.

30. Smit, S.K.; Eiben, A.E.; Szlávik, Z. An MOEA-based Method to Tune EA Parameters on Multiple Objective Functions. In Proceedings of the International Conference on Evolutionary Computation, (part of the International Joint Conference on Computational Intelligence IJCCI (ICEC)), Valencia, Spain, 24–26 October 2010; pp. 261–268.

31. Smit, S.K.; Eiben, A.E. Parameter Tuning of Evolutionary Algorithms: Generalist vs. Specialist. In *Applications of Evolutionary Computation*; Di Chio, C., Brabazon, A., Ebner, M., Farooq, M., Fink, A., Grahl, J., Greenfield, G., Machado, P., O'Neill, M., Tarantino, E., et al., Eds.; LNCS Springer: Berlin/Heidelberg, Germany, 2010; Volume 6024, pp. 542–551.

32. Branke, J.; Elomari, J.A. Meta-optimization for Parameter Tuning with a Flexible Computing Budget. In Proceedings of the Conference on Genetic and Evolutionary Computation Conference (GECCO), Philadelphia, PA, USA, 7–11 July 2012; pp. 1245–1252.

33. Eiben, A.E.; Smit, S.K. Parameter tuning for configuring and analyzing evolutionary algorithms. *Swarm Evol. Comput.* **2011**, *1*, 19–31. [CrossRef]

34. Blot, A.; Hoos, H.H.; Jourdan, L.; Kessaci-Marmion, M.É.; Trautmann, H. MO-ParamILS: A multi-objective automatic algorithm configuration framework. In Proceedings of the International Conference on Learning and Intelligent Optimization, Ischia, Italy, 29 May–1 June 2016; pp. 32–47.

35. Blot, A.; Pernet, A.; Jourdan, L.; Kessaci-Marmion, M.É.; Hoos, H.H. Automatically configuring multi-objective local search using multi-objective optimisation. In *Evolutionary Multi-Criterion Optimization*; Springer: Cham, Switzerland, 2017; pp. 61–76.

36. López-Ibánez, M.; Dubois-Lacoste, J.; Stützle, T.; Birattari, M. *The Irace Iackage, Iiterated Iace for Automatic Algorithm Configuration*; Technical Report TR/IRIDIA/2011-004; IRIDIA, Université Libre de Bruxelles: Bruxelles, Belgium, 2011.

37. Smit, S.K.; Eiben, A.E. Comparing Parameter Tuning Methods for Evolutionary Algorithms. In Proceedings of the IEEE Congress on Evolutionary Computation (CEC), Trondheim, Norway, 18–21 May 2009; pp. 399–406.

38. Liang, J.; Qu, B.; Suganthan, P. *Problem Definitions and Evaluation Criteria for the CEC 2013 Special Session on Real-Parameter Optimization*; Technical Report; Computational Intelligence Laboratory, Zhengzhou University: Zhengzhou, China; Nanyang Technological University: Singapore, 2013.

39. Zaharie, D. Critical values for the control parameters of Differential Evolution algorithms. In Proceedings of the 8th International Conference on Soft Computing, Brno, Czech Republic, 5–7 June 2002; pp. 62–67.

40. Ugolotti, R.; Cagnoni, S. Multi-objective Parameter Tuning for PSO-based Point Cloud Localization. In *Advances in Artificial Life and Evolutionary Computation. Proceedings of WIVACE 2014, Vietri sul Mare, Italy, 14–15 May 2014*; Pizzuti, C., Spezzano, G., Eds.; Communications in Computer and Information Science; Springer: Berlin/Heidelberg, Germany, 2014; Volume 445, pp. 75–85.

41. Ugolotti, R.; Mesejo, P.; Nashed, Y.S.G.; Cagnoni, S. GPU-Based Automatic Configuration of Differential Evolution: A Case Study. In *Progress in Artificial Intelligence*; Springer: Berlin/Heidelberg, Germany, 2013; pp. 114–125.

42. Kennedy, J.; Clerc, M. 2006. Available online: http://www.particleswarm.info/Standard_PSO_2006.c (accessed on 2 March 2019).

43. Montero, E.; Riff, M.C.; Pérez-Caceres, L.; Coello Coello, C. Are State-of-the-Art Fine-Tuning Algorithms Able to Detect a Dummy Parameter? In *Parallel Problem Solving from Nature Conference—PPSN XII, LNCS, Proceedings of the 12th International Conference, Taormina, Italy, 1–5 September 2012*; Coello, C.A., Cutello, V., Deb, K., Forrest, S., Nicosia, G., Pavone, M., Eds.; Springer: Berlin/Heidelberg, Germany, 2012; Volume 7491, pp. 306–315.

Article

Differential Evolution for Neural Networks Optimization

Marco Baioletti [1], Gabriele Di Bari [2], Alfredo Milani [1,*] and Valentina Poggioni [1]

[1] Department of Mathematics and Computer Science, University of Perugia, 06123 Perugia, Italy; marco.baioletti@unipg.it (M.B.); valentina.poggioni@unipg.it (V.P.)

[2] Department of Mathematics and Computer Science, University of Florence, 50100 Florence, Italy; gabriele.dibari@unifi.it

* Correspondence: alfredo.milani@unipg.it

Received: 5 November 2019; Accepted: 27 December 2019; Published: 2 January 2020

Abstract: In this paper, a Neural Networks optimizer based on Self-adaptive Differential Evolution is presented. This optimizer applies mutation and crossover operators in a new way, taking into account the structure of the network according to a per layer strategy. Moreover, a new crossover called *interm* is proposed, and a new self-adaptive version of DE called *MAB-ShaDE* is suggested to reduce the number of parameters. The framework has been tested on some well-known classification problems and a comparative study on the various combinations of self-adaptive methods, mutation, and crossover operators available in literature is performed. Experimental results show that DENN reaches good performances in terms of accuracy, better than or at least comparable with those obtained by backpropagation.

Keywords: neuroevolution; differential evolution; neural networks

---

1. Introduction

The use of Neural Network (NN) models has been steadily increasing in the recent past, following the introduction of Deep Learning methods and the ever-growing computational capabilities of modern machines. Thus, such models are applied to various problems, including image classification [1] and generation [2], text classification [3], speech recognition [4], emotion recognition [5], and many more. New and more complex network structures, such as Convolutional Neural Networks [6], Neural Turing Machines [7], and NRAM [8], were developed and applied to the aforementioned tasks; such new problems and structures also required the development of new optimization techniques [9–11].

According to these new trends, neuroevolution has also been renewed [12–15]. The term of neuroevolution is used to identify the research area where evolutionary algorithms are used to construct and train artificial neural networks. Several approaches have been proposed both to train networks' weights and topology and to exploit the characteristics of neuroevolution of being highly general, allowing learning with nondifferentiable activation functions, without explicit targets, and with recurrent networks [16,17].

The traditional method used by neural networks to learn their weights and biases is the gradient descent algorithm applied to a cost function and its most famous implementation is the backpropagation procedure. Nowadays, the backpropagation algorithm is still the workhorse of learning in neural networks even if its origin dates back to 1970s; its importance was revealed in 1986 [18].

Backpropagation works under two main assumptions about the form of the cost function: it has to be written as an average over cost functions C_x for individual training examples x and as a function of the outputs from the neural network. Moreover the activation functions have to be differentiable.

With that said, there are tricks for avoiding this kind of problems, and finding alternatives to gradient descent is an active area of investigation. An interesting analysis on the motivations according to backpropagation is the most used technique based on gradient to train neural networks and evolutionary approaches are not sufficiently studied is presented in [15].

As long as meta-heuristic algorithms are generally nondeterministic and not sensitive to the differentiability and continuity of the objective functions, these methods are used in a wide range of complex optimization problems. In addition, the stochastic global optimizations can identify global minimum without being trapped in local minima [19–21].

The most used evolutionary approach in neuroevolution is the genetic one, extensively employed in the conventional neuroevolution (CNE) [17,22] and also recently proposed in the case of deep neuroevolution [23]. In those algorithms, the best individuals (the individuals with the highest fitness) are evolved by means of the mutation and crossover operators and replace the genotypes with the lowest fitness in the population. The genetic approach is the most used technique because it is easy to implement and practical in many domains. However, on the other hand, there is the problem of the encoding since they use a discrete optimization method to solve continuous problems.

In order to avoid the encoding problem other continuous evolutionary meta-algorithms have been proposed including, in particular, differential evolution (DE). Indeed, DE evolves a population of real-valued vectors, so no encoding and decoding are required.

It is well known that DE performs better than other popular evolutionary algorithms [24], has a quick convergence, and is robust [25]; it also performs better for learning applications [26]. At the same time, DE has simple genetic operations, such as its operator of the mutation and survival strategy based on one-on-one competition. Moreover, they can also use population global information and individual local information to search for the optimal solution.

When the optimization problem is complex, the performance of the traditional DE algorithm depends on the selected the control parameters and mutation strategy [19,27–29]. If the control parameters and selected mutation strategy are unsuitable, then DE is likely to yield premature convergence, stagnation phenomena and excessive consumption of computational resources. In particular, the stagnation problem for DE applied to neural network optimization has been studied in [30].

In this paper the system DENN that optimizes artificial Neural Networks using DE is presented. The system uses a direct encoding with a one-to-one mapping between the weights of the neural networks and values of individuals in the population. This system is an enhanced version of the system introduced in [12], where a preliminary implementation was described.

A batching system is introduced to overcome one of the main computational problems of the proposed approach, i.e., the fitness computation. For every generation the population is evaluated on a limited number of training examples, given by the size of the current batch, rather than the whole training set. This reduces the computational load, particularly on large training sets. Moreover, a restart method is applied to avoid a premature convergence of the algorithm: the best individual is saved and the rest of the current population is discarded, continuing the research on a new random generated population.

Finally, a new self-adaptive mutation strategy *MAB-ShaDE* inspired to the multi-armed bandit UCB1 [31] and a new particular crossover operator *interm*, a randomized version of the arithmetic crossover, have been proposed.

An extensive experimental study have been implemented to (i) determine if this approach is scalable and applicable also to large classification problems, like MNIST digit recognition; (ii) study the performance reached by using *MAB-ShaDE* and *interm* components; and (iii) identify the best algorithm configurations, i.e., the configurations reaching the highest accuracy.

The experimental results show that DENN is able to outperform the backpropagation algorithm in training neural networks without hidden layers. Moreover, DENN is a viable solution also from a computational point of view, even if the time spent for learning is higher than its competitor BPG.

The paper is organized as follows. Background concepts about neuroevolution, DE algorithm and its self-adaptive strategies are summarized in Section 2, related works are presented in Section 3, the system is presented in Section 4, and experimental results are shown in Section 5. Section 6 closes the paper with some final considerations and some ideas for future works.

2. Background

2.1. Differential Evolution

Differential evolution (DE) is a evolutionary algorithm used for optimization over continuous spaces, which operates by improving a population of N candidate solutions evaluated by means of a fitness function f though a iterative process. The first phase is the initialization in which the first population is generated; there exists various approaches, among which the most common is randomly generating each vector. Following, during the iterative phase, for each generation a new population is computed though mutation and crossover operators; each new vector is evaluated and then the best ones are chosen, according to a selection operator, for the next generation. The evolution may proceed for a fixed number of generations or until a given criterion is met.

The mutation used in DE is called *differential mutation*. For each vector *target vector* x_i, for $i = 1, \ldots, N$, of the current generation, a vector $\bar{y}_i$, namely, *donor vector*, is calculated as linear combination of some vectors in the DE population selected according to a given strategy. In the literature, there exist many variants of the mutation operator (see for instance [32]). In this work, we implemented and used three operators: rand/1 [33], current_to_pbest [34], and DEGL [35].

The operator rand/1 is defined as

$$\bar{y}_i = x_a + F(x_b - x_c) \tag{1}$$

where $F \in [0, 2]$ is a real parameter called *mutation Factor*, a, b, c are unique random indices different from i.

The operator curr_to_pbest is defined as

$$\bar{y}_i = x_i + F(x_{pbest} - x_i) + F(x_a - x_b) \tag{2}$$

where $p \in (0, 1]$ and *pbest* is randomly selected index from the indices of the best $N \times p$ individuals of the population. Moreover, x_b is an individual randomly chosen from the set

$$\{x_1, \ldots, x_N\} \setminus \{x_a, x_i\} \cup \mathcal{A}$$

where $\mathcal{A}$ is an external archive of bounded size (usually with at most N individuals) that contains the individuals discarded by the selection operator.

Finally, DEGL is defined as

$$\begin{cases} \bar{y}_i = wL_i + (1 - w)G_i \\ L_i = x_i + \alpha(x_{nnbest} - x_i) + \beta(x_a - x_b) \\ G_i = x_i + \alpha(x_{best} - x_i) + \beta(x_a - x_b) \end{cases} \tag{3}$$

where *best* is the index of the best individual in the population, *nnbest* is the index of the best individual in the neighborhood of the target x_i, and $w \in [0, 1]$ is the weight of the convex combination between L_i and G_i.

The crossover operator creates a new vector y_i, namely *trial vector*, by recombining the donor with the corresponding target vector. There are many kinds of crossover; the most known is the binomial crossover where y_i is computed as follows,

$$y_{i,j} = \begin{cases} \bar{y}_{i,j} & \text{if } rand_{i,j} \leq CR \text{ or } j = j_{rand} \\ x_{i,j} & \text{otherwise} \end{cases} \quad \text{for } j = 1, \ldots, D \tag{4}$$

where $rand_{i,j} \in [0,1]$ is a real random number in $[0,1]$, j_{rand} is an integer random number in $\{1, \ldots, D\}$, and $CR \in [0,1]$ is the crossover probability.

Finally, the selection operator compares each trial vector y_i with the corresponding target vector x_i and selects the better of them in the population of the next generation.

2.1.1. Self-Adaptive Differential Evolution

The DE parameters F and CR have a strong impact during the evolution and the choose of their values is hard. In literature there exist many proposals of self-adaptive methods that select the values for F and CR.

One of the simplest and most popular method is *jDE* [36]. Each population individual x_i has its own values F_i and CR_i. The trial individual z_i inherits from the target the values F_i and CR_i, separately with probability 0.9; otherwise, a new value for F and/or for CR is randomly generated in $[0.1, 1]$ or in $[0, 1]$, respectively. The trial is then created using its own values for F and CR. If the trial survives in the selection phase, it will keep its values for F and CR in the next generation.

Another self-adaptive method is JADE [37], in which the value of F is randomly generated from a Cauchy distribution $C(\mu_F, 0.1)$ and the value of CR from the normal distribution $N(\mu_{CR}, 0.1)$. The means of these distributions μ_F and μ_{CR} are initialized to 0.5 and are updated at each generation as

$$\mu_F \leftarrow (1-c)\mu_F + cm_L(S_F)$$

and

$$\mu_{CR} \leftarrow (1-c)\mu_{CR} + cm_A(S_{CR})$$

where $m_L(S_F)$ is the Lehmer mean of the successful F values (i.e., those used to generate trials which are better than their targets) and $m_A(S_{CR})$ is the arithmetic mean of the successful CR values.

A variant of JADE is ShaDE [34], in which the values of F and CR are generated in the same way of JADE, but the means of the distributions are randomly selected from a *success history*, which stores the means computed with respect to the succesful trials.

Finally, L-ShaDE [38] is an enhancement of ShaDE where the population size is reduced as the generations go on.

2.1.2. Self-Adaptive Mutation

There also exist self-adaptive variants of DE which selects, for instance at each generation or even for each trial, the mutation operator to be applied among a set of possible choices.

We have decided to implement SaMDE [39]. It is a variant of *jDE*, where it is applied the automatic selection of mutation strategy from a pool of given strategies. Each population individual has its own vector V of o real numbers, where o is the number of mutation operators. The vector V is evolved in the same way as the individual itself. The values of V are used to randomly choose, by means of the roulette-wheel method, the mutation operator to be used to create the trial individual.

2.2. Neuroevolution

The term of neuroevolution is used to identify the research area where evolutionary algorithms are used to construct and train artificial neural networks. It covers a wide range of network architectures and neural models. Most neural learning methods focus on modifying the strengths of neural connections (i.e., their connection weights), whereas other models can optimize the structure of the network, the type of computation performed by individual neurons, and even learning rules that modify the network during evaluation.

The evolutionary approach dominating the scene of neuroevolution is the genetic approach by means of genetic algorithms. Typically, to find a network that solves the given task, a population of genetic encodings of neural networks (genotype) is evolved. The process constitutes an intelligent parallel search towards better genotypes in the space of solutions, and continues until a network with a sufficiently high fitness is found. The generate-and-test loop of evolutionary algorithms usually applied: (i) Each genotype is chosen in turn and decoded into the corresponding neural network, called phenotype. (ii) The performance of this network is then measured by a fitness value. (iii) After all individuals have been evaluated, genetic operators are applied and the next generation is created.

The evolution is applied to the the individuals with the highest fitness are crossed and mutated over with each other, and replace the genotypes with the lowest fitness in the population.

The conventional neuroevolution (CNE) follows this approach for the network weights [17,22]. This is the most used techniques because it is easy to implement but practical in many domains.

3. Related Works

The first DE-based optimizers for NNs were presented in the late '90s and the early 2000s by [40,41], who presented and analyzed the applications of DE on the problem of feedforward NN train. In recent times, new applications of evolutionary algorithms have been presented in the area of neuroevolution [32].

The dominating evolutionary approach used is the genetic one [17,22]: this is used to optimize both topology and weights of the network but in the latter case it is very limited by being a discrete approach. In literature several encodings for the real weights are proposed, with genes represented either as a real-valued string or characters sequence, which can be interpreted as real values with a specific precision using for example Gray-coded numbers.

More adaptive approaches have been suggested, for example in [42] or more recently in [43]. In the first paper, the authors presented a dynamic encoding, which depends on the exploitation and exploration phases of the search. In the second one, the authors proposed a self-adaptive encoding, where the string characters are interpreted as a system of particles whose center of mass determines the encoded value. Other adaptive approaches have been developed for network immunization and diffusion in link prediction [44,45].

Moreover, they have also used a direct encoding that exploits the particular problem structure.

These methods are not general and are not easily extendable to be applicable in more general cases [17]. In [46], a direct encoding floating-point representation of the NN's weights is used. Precisely, the authors use the evolution strategy called CMA-ES, a real-value optimization algorithm, applied to the well-known reinforcement learning problem: pole balancing.

Among DE applications to neuroevolution, the most related works we have to cite are [13–15,30,47], even if they apply the evolutionary meta-heuristics in a different way.

In [47], the search exploration is enhanced by a DE algorithm with a modified best mutation operation: the algorithm is used to train the network and the global best value is used as a seed by the backpropagation procedure (BPG).

In [13], three different methods (GA, DE, and EDA) are compared and used to train a simple network architecture with one hidden layer, the learning factor, and the seed for the weights initialization.

In [14], the authors use the Adaptive DE (ADE) algorithm to calculate the initial weights and the thresholds of standard neural networks trained by BPG. The authors demonstrated that the system is effective to solve time series forecasting problems.

In [15], a Limited Evaluation Evolutionary Algorithm (LEEA) is applied to optimize the weights of the network. This paper is related to our paper because we employ a similar batching system, in which minibatches are used in the training phase and are changed after a certain number of generations.

The work in [30] has a strong connection with ours because the author studied how different mutation operators work to train neural networks. The results showed that the DEGL-trig (a composition of DEGL with Trigonometric mutation) is the best mutation operator to use with small NNs.

DE and the other enhancement methods permit our algorithm to train neural networks much larger than those used in [15,30]: whereas the maximum size handled in [15] has less than 1500 weights and the maximum size handled in [30] has only 46 weights, we are capable to train a feedforward neural network for MNIST which has more than 7000 weights.

4. The DENN Algorithm

This section describes the Differential Evolution for Neural Networks. The idea is to apply the Differential Evolution for optimization of NN's weights taking in count the structure of the network.

Given a fixed topology and fixed activation functions, a population $\mathcal{P}$ is defined as a set of N neural networks.

We decided to exploit the DE characteristic of working with continuous values by using a direct codification based on a one-to-one mapping between the weights of the neural network and individuals in DE population.

More precisely, let N be a feedforward neural network composed of L levels. For each level, l, of the network is defined by a real valued matrix, $\mathbf{W}^{(l)}$, and a real valued vector, $\mathbf{b}^{(l)}$, representing, respectively, the connection weights and the bias values. Therefore, each population individual x_i is defined as a sequence

$$\langle (\hat{\mathbf{W}}^{(i,1)}, \mathbf{b}^{(i,1)}), \ldots, (\hat{\mathbf{W}}^{(i,L)}, \mathbf{b}^{(i,L)}) \rangle,$$

where $\hat{\mathbf{W}}^{(i,l)}$ is the real values vector obtained by linearization of the matrix $\mathbf{W}^{(i,l)}$, for $l = 1, \ldots, L$.

For a population individual x_i, we indicate by $x_i^{(h)}$ its h–th component, for $h = 1, \ldots, 2L$. For example, $x_i^{(h)} = \hat{\mathbf{W}}^{(i,(h+1)/2)}$, if h is odd, whereas $x_i^{(h)} = \mathbf{b}^{(i,h/2)}$ if h is even.

Note that for each solution x_i the component $x_i^{(h)}$ is a vector whose size $d^{(h)}$ is dependent on the number of neurons of in the level h.

The individuals of the population are evolved by applying mutation and crossover operators in a component-wise way. For instance, the mutation $rand/1$ for the individual x_i is applied as three indices, a, b, c, that are randomly chosen in the set $\{1, \ldots, N\} \setminus \{i\}$ without repetition; then, for $h = 1, \ldots, 2L$, the h–th component $\bar{y}_i^{(h)}$ of the donor individual $\bar{y}_i$ is calculated as the linear combination

$$\bar{y}_i^{(h)} = x_a^{(h)} + F(x_b^{(h)} - x_c^{(h)}).$$

The evaluation of a population element x_i is performed by a fitness function f, which is the objective function to be optimized.

As proposed in the many other efficient applications, we split the dataset D in three different subsets: a training set TS, a validation set VS, and a test set ES. The TS is used for the training phase, the VS is used at the end of each training phase for a uniform evaluation of the individuals, and ES is used on the best neural network in order to evaluate the performance.

As the evaluation phase is the most time consuming operation, and it can lead to unacceptable computation time if the fitness is computed on the whole dataset, we decided to use a batching method similar to the one proposed in [15] by partitioning the training set TS in k batches $B_0, \ldots, B_{k-1}$ of size $b = |TS|/k$.

Note that records in each batch should follow the same distribution to avoid the risk of the overfitting, followed by generation of a model that is unable to generalize.

At each generation the population is evaluated against only a small number of training samples, given by the size of the current batch, instead of evaluating the population with all the training set samples. This permits to reduce the computational load, especially on large training sets.

To reduce the problems that arose when the batch is changed as well as obtaining a smoother transition from a batch to the next one, we defined a window U of size b, which is a set of samples taken from the current batch B_i and from the next one B_{i+1}.

At the beginning of an epoch, the fitness of all individuals in $\mathcal{P}$ is re-evaluated by computing the fitness on the new batch defined by currently window U.

The window is changed after s generations, by substituting b/r examples of U from B_i with b/r examples taken from B_{i+1} and not already present in U.

Then, given *sub-epoch* dimension s, the window passes from a batch to the next one in r *sub-epoch*, or in other words in rs generations (we call *epoch* this period). In this way, the fitness function change more smoothly and the evolution has more time to learn from the batch because the window is updated after s generations.

Moreover, the batches are reused in a cyclic way; when the algorithm iterates for more than k epochs and thus runs out of available batches, the batch sequence restarts from the first one.

Since the fitness function relies also on the batch and we need a fixed one to compare the individuals across the *epochs*; consequently, at the end of every epoch e, the best individual x_e^* is calculated as the NN in $\mathcal{P}$, which reaches the highest accuracy in the validation set VS. The global best network x^{**} found so far is then eventually updated.

A restart method is used to avoid a premature convergence of the algorithm; The restart strategy adopted discard all the individuals in the current population, except the best one, and for the next algorithm iteration a new population randomly generated is used. The restart technique is applied at the end of each epoch e, if the fitness evaluation of x^{**} did not change for a given number M of epochs. The complete algorithm, namely DENN, is depicted in Algorithm 1.

Algorithm 1: The algorithm DENN

Initialize the population;
$k \leftarrow |TS|/b$;
Extract the k batches $B_0, \ldots, B_{k-1}$;
$h \leftarrow 0$;
for $e \leftarrow 0$ **to** $G/(rs) - 1$ **do**
    Set the current batch as $B_{e \bmod k}$;
    Re-evaluate all the elements $(x_1, \ldots, x_N)$;
    for $z \leftarrow 1$ **to** r **do**
        for $j \leftarrow 1$ **to** s **do**
            for $i \leftarrow 1$ **to** N **do**
                $y_i \leftarrow$ generate_offspring(x_i);
                Evaluate the fitness $f(y_i)$
            for $i \leftarrow 1$ **to** N **do**
                if $f(y_i) > f(x_i)$ **then**
                    $x_i \leftarrow y_i$
        Update the window U
    $x_e^*, f_e^* \leftarrow$ best_score$(x_1, \ldots, x_N)$;
    Update x^{**}, f^{**};
    if x^{**} *is not changed* **then**
        if $h > M$ **then**
            Restart the population;
            $h \leftarrow 0$;
    else
        $h \leftarrow h + 1$
return x^{**};

In the algorithm DENN, the function *generate_offspring* execute the mutation and the crossover operators in order to produce the *trial individual*, whereas the function *best_score* finds the best network x^* and computes the respective score f^* among all the individuals in the population.

4.1. Fitness Function

In the case of classification problems, the fitness function used to evaluate the individual x is the well-known cross-entropy. In this case, the optimization problem is to find the neural network x minimizing the $H(x)$ value, computed as

$$H(x) = - \sum_{i=1}^{b} \sum_{j=1}^{C} z'_{ij} \log(z_{ij}) \tag{5}$$

where z'_{ij} and z_{ij} are, respectively, the value predicted by x and the actual value for the i-th record of U with respect to the j-th class (C is the number of classes).

4.2. The Interm Crossover

We have implemented a new particular crossover operator called *interm*, which is a randomized version of the arithmetic crossover. If x_i is the target and $\bar{y}_i$ is the donor, then the trial y_i is obtained in the following way; for each component $x_i^{(h)}$ of x_i and $\bar{y}_i^{(h)}$ of $\bar{y}_i$, let $a_i^{(h)}$ be a vector of $d^{(h)}$ randomly numbers, generated with a uniform distribution $[0,1]$, then

$$y_{ij}^{(h)} = a_{ij}^{(h)} x_i^{(h)} + (1 - a_{ij}^{(h)}) \bar{y}_{ij}^{(h)}$$

for $j = 1, \ldots, d^{(h)}$.

4.3. The MAB-ShaDE Mutation Method

We have also implemented a variant of ShaDE algorithm, called MAB-ShaDE. MAB-ShaDE has a solution archive and a history of the best CR and F parameters, like ShaDE (Section 2).

The novelty of MAB-ShaDE is in the method used, inspired to the Multi-armed bandit UCB1 [31], to select one mutation strategy among a list of possible operators.

We consider the mutation strategies as arms of the bandit and the epochs as the rounds where the reward of the selected arm is computed. Therefore, for each mutation operator OP, UCB1 stores the average value of the reward μ_{OP} and the number of epoch n_{OP} in which OP has been used. After the end of the epoch e, the operator

$$O = \arg \max_{OP} \left(\mu_{OP} + \sqrt{2 \log e / n_{OP}} \right)$$

is chosen as mutation strategy for the next epoch.

5. Experiments

In this section, we describe the experiments performed to assess the effectiveness of DENN algorithm as an alternative to backpropagation for neural network optimization.

Moreover, we are interested to find the best algorithm combination and, in particular, the best mutation and crossover operators. To do that we organized two rounds of experiments. First of all, we tested all the possible combinations in order to define the best algorithm singularly for each dataset and the global best. These experiments are described in Section 5.3 and allow us to conclude that there is no winner combination if we consider the results grouped by dataset, whereas we can say that the combination of ShaDE with *curr_p_best* and *interm* globally perform better than any other combination. Then, we decided to verify the effectiveness both in term of computational effort and accuracy compared to the classical backpropagation. These results are shown in Section 5.2.

All the networks used in these experiments are without any hidden layer.

DENN has been implemented as a C++ program (Source code available at https://github.com/Gabriele91/DENN). The results presented here are obtained with a computer having a CPU AMD Ryzen 1600 and 16GB RAM.

5.1. Datasets

We tested DENN on various classification datasets from the UCI repositories (https://archive.ics.uci.edu/ml/datasets) (MAGIC, QSAR, and GASS) and also on the well-known MNIST (http://yann.lecun.com/exdb/mnist/) dataset for hand-written digit classification. They have been chosen because of their differences on the number of features and records. Moreover, we chose the MNIST dataset because it is a classical challenge with well-known results obtained by various NN classification systems. Note that these datasets are also considered as interesting challenges in [15].

- MAGIC Gamma telescope:dataset with 19,020 records, 10 features, and two classes.
- QSAR biodegradation: dataset with 1055 records, 41 features, and two classes.
- GASS Sensor Array Drift: dataset with 13,910 records, 128 features, and six classes.
- MNIST: dataset with 70,000 records, 784 features, and 10 classes.

5.2. System Parameters

The DENN algorithm depends on various parameters: some directly deriving from the DE (F, CR, the auto-adaptive variant of DE, the mutation, and crossover operators), other depending on the batching system (s, b, and r). For each dataset we analyzed the following parameters,

- the auto-adaptive variant of DE (simply called *Method*),
- the *Mutation* operator,
- the *Crossover* operator,
- the number s of generations of a sub-epoch,
- the batch/window size b, and
- the ratio r between the batch size and the number of records changed in the window at each sub-epoch.

and their values are shown in Table 1.

Table 1. Parameters values.

Parameter	Values
Method	*JDE, JADE, ShaDE, L-ShaDE, MAB-ShaDE, SAMDE*
Mutation	*rand/1, curr_to_pbest, DEGL*
Crossover	*bin, interm*
b	*low, mid, high*
r	$1, \frac{1}{2}, \frac{1}{4}$
s	$\frac{b}{r}, \frac{b}{2r}, \frac{b}{4r}$

We have chosen three levels for the window size b, called *low*, *mid*, and *high*, which depend on the dataset size, hence they correspond to different values for each dataset (see Table 2).

Table 2. Size of batches for each dataset.

Dataset	*Low*	*Mid*	*High*
MAGIC	20	40	80
QSAR	10	20	40
GASS	20	40	80
MNIST	50	100	200

We have also chosen three levels for the length s of the sub-epoch, which are proportional to the number b/r of records changed at each sub-epoch. For instance, the lowest level is $\frac{b}{4r}$, which corresponds to a number of generations equal to $1/4$ of b/r. The main motivation of this choice is that DENN should need more generations with larger batches/windows.

Another aspect of our tests is that we have used a double version for each dataset, the original one and the normalized one. In this way, we can see if the normalization process affects the performances of DENN.

As we implemented a complete test for each possible combination in each dataset and we run the same configuration five times, we collected accuracy values and computation time for 30,240 runs.

All the results are stored on GitHUB (Results available at https://github.com/Gabriele91/DENN-RESULTS-2019); in this paper, only the most significant are shown.

5.3. Algorithm Combination Analysis

The first analysis has been made on the convergence graphics, where for each dataset the data of accuracy has been plotted during the generations. For each dataset and for each self-adaptive method, the data of the method which obtained the highest accuracy have been displayed in Figures 1–4.

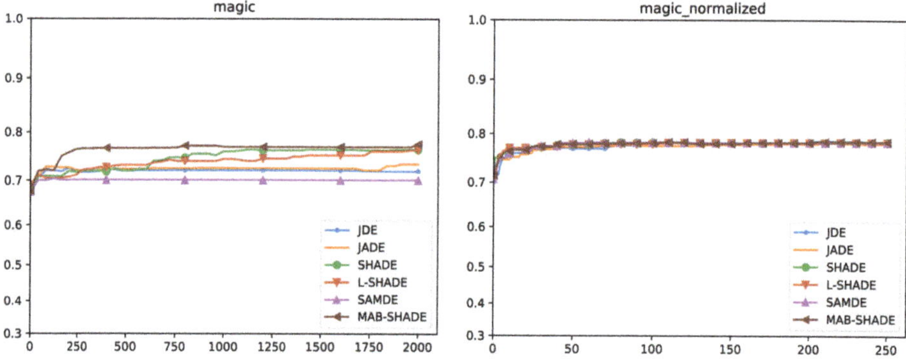

Figure 1. Plots of convergences on MAGIC and MAGIC normalized datasets.

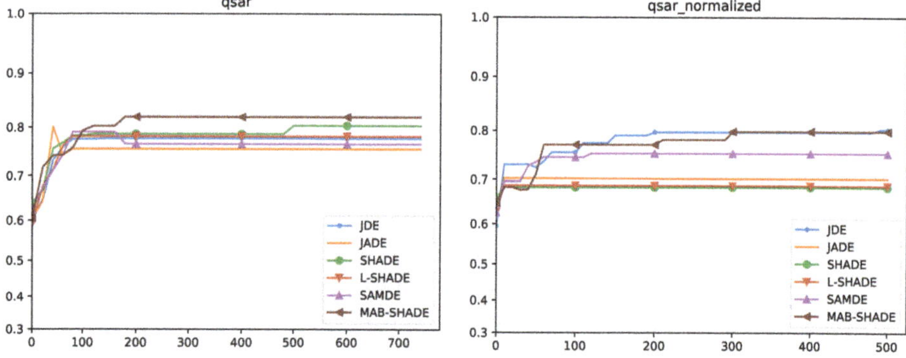

Figure 2. Plots of convergences on QSAR and QSAR normalized datasets.

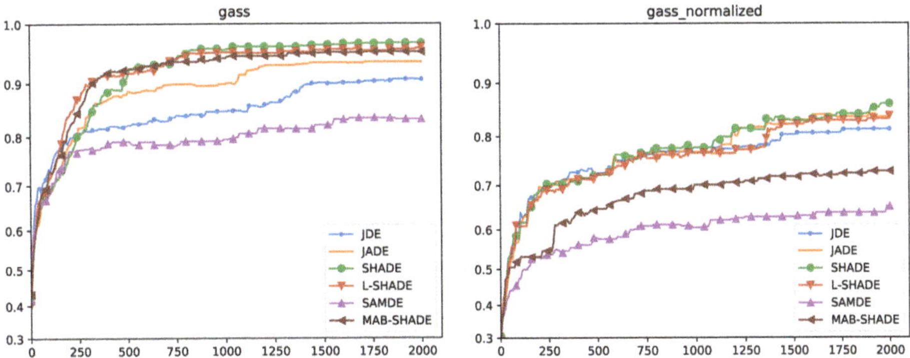

Figure 3. Plots of convergences on GASS and GASS normalized datasets.

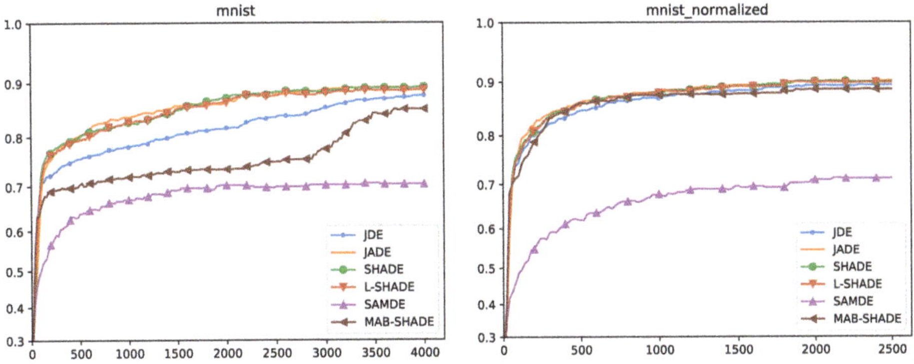

Figure 4. Plots of convergences on MNIST and MNIST normalized datasets.

From the plots, it is possible to see that, excluding the cases where the differences are not significant, *MAB-ShaDE* works well on smaller datasets (MAGIC and QSAR), whereas ShaDE is the best method for larger datasets.

5.4. Convergence Analysis

In this subsection, we discuss the convergence across all DE used and analyzed in this paper on the datasets discussed before. On the MAGIC dataset, *SHADE* and *L-SHADE* converge in around 1750 generations, whereas the proposed *MAB-SHADE* requires only 250 generations to achieve a solution with a comparable quality. Other methods were able to discover lower quality solutions only. Regarding the other binary classification problem, QSAR, *MAB-SHADE* converges faster than all the other methods in less than 200 generations, while simultaneously obtaining a higher quality solution.

On the GASS multi-class problem, *MAB-SHADE* follow the same convergence path of *L-SHADE*, whereas *SHADE* has a slow convergence, but the quality of result reached by *SHADE* is slightly better, conversely, the other methods do not reach a satisfactory solution.

On the image classification problem MNIST, *SHADE* and *L-SHADE* resulted the best algorithms in terms of the solution quality and the time of convergence, whereas the other methods did not obtain comparable solutions in terms of quality; noticeably, *MAB-SHADE* did not get stuck, but it is likely that it requires more generations to converge to a solution.

We also performed the same tests with normalized versions of the datasets, finding susceptible differences with the previous results.

On MAGIC all the methods converged to the same solution, whereas on QSAR the best solution was reached by *MAB-SHADE* and *jDE*. Regarding GASS, no analyzed method reached a solution comparable in terms of quality to the solutions found on the corresponding original dataset. Finally, on MNIST all the methods, except *SAMDE*, reached good solutions, which are, however, below the solutions found with *SHADE* on the non normalized datasets.

Generally speaking, the best DE method is *SHADE* for the multi-class problems and *MAB-SHADE* for binary classification. Anyway, the convergence curves of *SHADE* are close to those of *MAB-SHADE* in the latter kind of problems.

Finally, it is worth to notice that *MAB-SHADE* performed systematically better than its direct competitor *SAMDE* without requiring to choose a particular mutation strategy.

Quade Weighted Rank

As we have different results for different datasets, we applied the Quade test [48] in order to obtain a global ranking which takes into account the differences among the datasets.

The Quade test considers that some datasets could be more difficult to deal with (i.e., the differences in accuracy of the various algorithms are larger). In this way, the rankings computed on each dataset are scaled depending on the differences observed in the algorithms' performances [48].

With reference to Table 3, for each algorithm combination, the weighted ranking values are shown in the last column *Quade rank*.

These values are computed as follows.

Given the 756 parameter configurations we obtained by varying the values for each dimensions as shown in Table 1, we memorized in v_{ij} the average accuracy value obtained by the configuration in the row i on the dataset in the column j. The ranks r_{ij} of these values are computed for each dataset. Ranks are also assigned to the datasets according to the sample range of accuracy values obtained on it. The sample range within data set j is the difference between the largest and the smallest accuracy v_{ij} within that data set. Let Q_j be the rank assigned to the j-th dataset with respect to these values. Then, the Quade weighted rank is obtained ordering the parameters configuration with respect to $S_j = \sum_i r_{ij} Q_j$.

In Table 3, the top 20 among the 756 configurations tested are shown. We can see that *SHADE*, *curr_p_best*, *interm*, and $b = high$ are the best choices.

Table 3. Top 20 Quade ranking for parameter configurations.

Rank	Method	Mutation	Crossover	b	r	s	QUADE Rank
1	SHADE	*curr_p_best*	*interm*	high	1	$\frac{B}{2R}$	2713
2	SHADE	*curr_p_best*	*interm*	high	2	$\frac{B}{4R}$	3011
3	L–SHADE	*curr_p_best*	*interm*	high	4	$\frac{B}{R}$	3199
4	SHADE	*curr_p_best*	*interm*	mid	4	$\frac{B}{4R}$	3291
5	SHADE	*curr_p_best*	*interm*	high	1	$\frac{B}{R}$	3330
6	SHADE	*curr_p_best*	*interm*	low	2	$\frac{B}{2R}$	3365
7	JDE	*degl*	*bin*	high	2	$\frac{B}{4R}$	3721
8	JADE	*curr_p_best*	*interm*	high	2	$\frac{B}{R}$	3808
9	JDE	*curr_p_best*	*interm*	high	1	$\frac{B}{R}$	3838
10	L–SHADE	*curr_p_best*	*interm*	high	1	$\frac{B}{2R}$	3872
11	L–SHADE	*curr_p_best*	*bin*	high	2	$\frac{B}{2R}$	3897
12	SHADE	*curr_p_best*	*interm*	high	4	$\frac{B}{4R}$	3928
13	SHADE	rand/1	*interm*	high	1	$\frac{B}{2R}$	4108
14	L–SHADE	*curr_p_best*	*interm*	high	2	$\frac{B}{R}$	4160
15	MAB–SHADE	None	*bin*	high	1	$\frac{B}{R}$	4194
16	JADE	*curr_p_best*	*interm*	mid	4	$\frac{B}{2R}$	4267
17	SHADE	*curr_p_best*	*interm*	high	1	$\frac{B}{4R}$	4355
18	SHADE	*curr_p_best*	*bin*	high	2	$\frac{B}{2R}$	4356
19	JADE	*degl*	*bin*	low	2	$\frac{B}{4R}$	4423
20	JDE	*degl*	*bin*	high	4	$\frac{B}{2R}$	4513

5.5. Execution Times

The execution time of DENN changes with respect to the number of features and the size of batch. Therefore, in Table 4, we show the average execution time in seconds of DENN in each datasets and for each level of b. Note that the execution time is not sensitively affected by the normalization of the datasets.

Table 4. Average execution times.

Dataset	b = Low	b = Mid	b = High
MAGIC	0.571	0.593	0.612
MAGIC-N	0.583	0.591	0.625
QSAR	0.760	0.776	0.783
QSAR-N	0.748	0.752	0.779
GASS	7.740	9.676	10.819
GASS-N	7.751	9.617	10.797
MNIST	133.748	154.731	194.948
MNIST-N	132.615	155.842	191.055

In Table 4, the worst case required approximately three minutes for the computation of the solution also thanks to a strong parallelization of the computation. Note that this point is a plus of the evolutionary approach: in the case of an iterative method like backpropagation it would have been impossible. Therefore, we can conclude that the time to reach the solution is reasonable and the approach is feasible, even if it is slower when compared to gradient-based methods.

5.6. Comparison with Backpropagation

In this section, we compared our method to the Backpropagation (BPG) algorithm, using two optimizer: the Stochastic Gradient Descendent (SGD) and the more powerful Adam. The experiments were performed on the same datasets MAGIC, QSAR, GASS, and MNIST, using both the original and the normalized versions.

The results are reported in Table 5, where for each dataset we compare the classification accuracy obtained by NNs trained with BPG (using both optimizers) to the accuracy obtained by our method (DENN). As it can be seen in the results, in such a scenario our method shows better performances or, in some cases, comparable to the competitors. More specifically, DENN obtained higher accuracy if compared to SGD on all classification problems, while ADAM performed better only on MNIST.

Table 5. Comparison BPG - DENN.

Dataset	BPG+SGD	BPG+ADAM	DENN
MAGIC	73.0%	71.9%	76.6%
MAGIC-N	77.0%	78.2%	78.8%
QSAR	78.9%	72.5%	79.9%
QSAR-N	80.6%	80.6%	81.4%
GASS	73.9%	68.9%	86.2%
GASS-N	89.8%	94.6%	96.8%
MNIST	90.3%	90.7%	89.4%
MNIST-N	90.1%	90.5%	90.4%

The difference between MNIST and other datasets is about their features. In MNIST, the features are just quantitative; whereas, in the other ones, some data has a quantitative nature and other data are qualitative.

Generally, all the algorithms work better on normalized datasets, except that in MNIST, where data have are already a high degree of homogeneity. On the other hand, in GASS the effect of using normalized datasets is much greater for all the algorithms.

Note that our method can be useful in MLP networks trained for problems on which traditional algorithms can hardly achieve satisfying performances or need larger networks to achieve the same results.

6. Conclusions and Future Works

In this paper the DENN framework, a learning algorithm for Neural Networks based on Self-Adaptive Differential Evolution, is presented. Experiments show that the framework is able to solve classification problems, reaching satisfying levels of accuracy even in case of large datasets. The use of batch systems allows the application of DE to new untested domains. Indeed, it is worth noticing that the size of the problems handled in this work is significantly larger than those tested in other works available in literature.

Furthermore, the per-layer mutation and crossover strategies introduced in this work perform better than the traditional DE used in previous works. From the experiments we found the following:

- the configuration of the Self-Adaptive *ShaDE* with *curr_p_best* and the new *interm* crossover performs better than other settings,
- the slow change of batches allows to reach better results, and
- the MAB-ShaDE algorithm reduces the number of parameters at the cost of slightly worse solutions.

The results obtained with DENN are almost always better than those obtained with backpropagation. Moreover DENN appears to be robust than its competitor with respect to the normalization.

Future research will investigate the possibility of using DENN as optimizer for other Neural Network structures, including Convolutional Neural Networks, Recurrent Neural Networks, and Neural Turing Machines. Another scenario could be the application of Evolutionary Algorithms to those problems and domains where gradient-based optimizers do not perform as well as in supervised learning. A first direction will be the application of DENN in the Reinforcement Learning context, where a NN approximates the Value-Action Function (or Q Function) for agents in a nonlinear and complex environment.

Author Contributions: Writing-original draft: A.M., M.B., V.P., G.D.B.; Conceptualization and all other contributions: A.M., M.B., V.P., G.D.B. All authors have read and agreed to the published version of the manuscript.

Funding: This research has been partially supported by *Progetti Ricerca di Base 2015–2019 Baioletti-Milani-Poggioni* granted by Department of Mathematics and Computer Science University of Perugia, Italy.

References

1. Cireşan, D.C.; Meier, U.; Gambardella, L.M.; Schmidhuber, J. Deep, big, simple neural nets for handwritten digit recognition. *Neural Comput.* **2010**, *22*, 3207–3220. [CrossRef] [PubMed]
2. Goodfellow, I.; Pouget-Abadie, J.; Mirza, M.; Xu, B.; Warde-Farley, D.; Ozair, S.; Courville, A.; Bengio, Y. Generative adversarial nets. In *Advances in Neural Information Processing Systems*; MIT: Cambridge, MA, USA, 2014; pp. 2672–2680.
3. Santucci, V.; Spina, S.; Milani, A.; Biondi, G.; Di Bari, G. Detecting hate speech for Italian language in social media. In Proceedings of the EVALITA 2018, co-located with the Fifth Italian Conference on Computational Linguistics (CLiC-it 2018), Turin, Italy, 12–13 December 2018; Volume 2263.
4. Graves, A.; Jaitly, N.; Mohamed, A.R. Hybrid speech recognition with deep bidirectional LSTM. In Proceedings of the 2013 IEEE workshop on automatic speech recognition and understanding, Olomouc, Czech Republic, 8–12 December 2013; pp. 273–278.
5. Biondi, G.; Franzoni, V.; Poggioni, V. A deep learning semantic approach to emotion recognition using the IBM watson bluemix alchemy language. In *Lecture Notes in Computer Science*; Springer: Cham, Switzerland, 2017; pp. 719–729.

6. Krizhevsky, A.; Sutskever, I.; Hinton, G.E. Imagenet classification with deep convolutional neural networks. In *Advances in Neural Information Processing Systems*; MIT: Cambridge, MA, USA, 2012; pp. 1097–1105.

7. Graves, A.; Wayne, G.; Danihelka, I. Neural turing machines. *arXiv* **2014**, arXiv:1410.5401.

8. Kurach, K.; Andrychowicz, M.; Sutskever, I. Neural Random-Access Machines. *arXiv* **2015**, arXiv:abs/1511.06392.

9. Hinton, G.; Deng, L.; Yu, D.; Dahl, G.; Mohamed, A.R.; Jaitly, N.; Senior, A.; Vanhoucke, V.; Nguyen, P.; Kingsbury, B.; et al. Deep Neural Networks for Acoustic Modeling in Speech Recognition: The Shared Views of Four Research Groups. *IEEE Signal Process. Mag.* **2012**, *29*, 82–97. [CrossRef]

10. Krizhevsky, A.; Sutskever, I.; Hinton, G.E. ImageNet Classification with Deep Convolutional Neural Networks. In *Advances in Neural Information Processing Systems 25*; Pereira, F., Burges, C.J.C., Bottou, L., Weinberger, K.Q., Eds.; Curran Associates, Inc.: Dutchess County, NY, USA, 2012; pp. 1097–1105.

11. Bengio, Y.; Goodfellow, I.J.; Courville, A. Deep learning. *Nature* **2015**, *521*, 436–444.

12. Baioletti, M.; Di Bari, G.; Poggioni, V.; Tracolli, M. Can Differential Evolution Be an Efficient Engine to Optimize Neural Networks? In *Machine Learning, Optimization, and Big Data*; Springer International Publishing: Cham, Switzerland, 2018; pp. 401–413.

13. Donate, J.P.; Li, X.; Sánchez, G.G.; de Miguel, A.S. Time series forecasting by evolving artificial neural networks with genetic algorithms, differential evolution and estimation of distribution algorithm. *Neural Comput. Appl.* **2013**, *22*, 11–20. [CrossRef]

14. Wang, L.; Zeng, Y.; Chen, T. Back propagation neural network with adaptive differential evolution algorithm for time series forecasting. *Expert Syst. Appl.* **2015**, *42*, 855–863. [CrossRef]

15. Morse, G.; Stanley, K.O. Simple Evolutionary Optimization Can Rival Stochastic Gradient Descent in Neural Networks. In Proceedings of the Genetic and Evolutionary Computation Conference (GECCO), Denver, CO, USA, 20–24 July 2016; ACM: New York, NY, USA, 2016; pp. 477–484.

16. Miikkulainen, R. Neuroevolution. In *Encyclopedia of Machine Learning*; Springer: Boston, MA, USA, 2010; pp. 716–720.

17. Floreano, D.; Dürr, P.; Mattiussi, C. Neuroevolution: from architectures to learning. *Evol. Intell.* **2008**, *1*, 47–62. [CrossRef]

18. Rumelhart, D.E.; Hinton, G.E.; Williams, R.J. Learning representations by back-propagating errors. *Nature* **1986**, *323*, 533. [CrossRef]

19. Eltaeib, T.; Mahmood, A. Differential Evolution: A Survey and Analysis. *Appl. Sci.* **2018**, *8*, 1945. [CrossRef]

20. Kitayama, S.; Arakawa, M.; Yamazaki, K. Differential evolution as the global optimization technique and its application to structural optimization. *Appl. Soft Comput.* **2011**, *11*, 3792–3803. [CrossRef]

21. Zou, D.; Wu, J.; Gao, L.; Li, S. A modified differential evolution algorithm for unconstrained optimization problems. *Neurocomputing* **2013**, *120*, 469–481. [CrossRef]

22. Yao, X. Evolving artificial neural networks. *Proc. IEEE* **1999**, *87*, 1423–1447.

23. Such, F.P.; Madhavan, V.; Conti, E.; Lehman, J.; Stanley, K.O.; Clune, J. Deep Neuroevolution: Genetic Algorithms Are a Competitive Alternative for Training Deep Neural Networks for Reinforcement Learning. *arXiv* **2017**, arXiv:1712.06567.

24. Vesterstrom, J.; Thomsen, R. A comparative study of differential evolution, particle swarm optimization, and evolutionary algorithms on numerical benchmark problems. In Proceedings of the 2004 Congress on Evolutionary Computation (IEEE Cat. No.04TH8753), Portland, OR, USA, 19–23 June 2004; Volume 2, pp. 1980–1987.

25. Price, K.; Storn, R.M.; Lampinen, J.A. *Differential Evolution: A Practical Approach to Global Optimization*; Springer Science & Business Media: Berlin/Heidelberg, Germany, 2006.

26. Zhang, X.; Xue, Y.; Lu, X.; Jia, S. Differential-Evolution-Based Coevolution Ant Colony Optimization Algorithm for Bayesian Network Structure Learning. *Algorithms* **2018**, *11*, 188. [CrossRef]

27. Gämperle, R.; Müller, S.D.; Koumoutsakos, P. A parameter study for differential evolution. *Adv. Intell. Syst. Fuzzy Syst. Evol. Comput.* **2002**, *10*, 293–298.

28. Santucci, V. Linear Ordering Optimization with a Combinatorial Differential Evolution. In Proceedings of the IEEE International Conference on Systems, Man, and Cybernetics, SMC 2015, Kowloon, China, 9–12 October 2015; IEEE Press: Piscataway, NJ, USA, 2016; pp. 2135–2140, doi:10.1109/SMC.2015.373.

29. Santucci, V. A differential evolution algorithm for the permutation flowshop scheduling problem with total flow time criterion. In *Lecture Notes in Computer Science*; LNCS 8672; Springer: Berlin/Heidelberg, Germany, 2016; pp. 161–170, ISSN 03029743.

30. Piotrowski, A.P. Differential Evolution algorithms applied to Neural Network training suffer from stagnation. *Appl. Soft Comput.* **2014**, *21*, 382–406. [CrossRef]

31. Chen, W.; Wang, Y.; Yuan, Y.; Wang, Q. Combinatorial Multi-armed Bandit and Its Extension to Probabilistically Triggered Arms. *J. Mach. Learn. Res.* **2016**, *17*, 1746–1778.

32. Das, S.; Mullick, S.S.; Suganthan, P. Recent advances in differential evolution—An updated survey. *Swarm Evol. Comput.* **2016**, *27*, 1–30. [CrossRef]

33. Storn, R.; Price, K. Differential Evolution—A Simple and Efficient Heuristic for global Optimization over Continuous Spaces. *J. Glob. Optim.* **1997**, *11*, 341–359.:1008202821328. [CrossRef]

34. Tanabe, R.; Fukunaga, A. Success-History Based Parameter Adaptation for Differential Evolution. In Proceedings of the IEEE Congress on Evolutionary Computation, Cancun, Mexico, 20–23 June 2013.

35. Das, S.; Abraham, A.; Chakraborty, U.K.; Konar, A. Differential Evolution Using a Neighborhood-Based Mutation Operator. *IEEE Trans. Evol. Comput.* **2009**, *13*, 526–553. [CrossRef]

36. Brest, J.; Boskovic, B.; Mernik, M.; Zumer, V. Self-Adapting Control Parameters in Differential Evolution: A Comparative Study on Numerical Benchmark Problems. *IEEE Trans. Evol. Comput.* **2006**, *10*, 646–657. [CrossRef]

37. Peng, F.; Tang, K.; Chen, G.; Yao, X. Multi-start JADE with knowledge transfer for numerical optimization. In Proceedings of the 2009 IEEE Congress on Evolutionary Computation, Trondheim, Norway, 18–21 May 2009; pp. 1889–1895.

38. Tanabe, R.; Fukunaga, A.S. Improving the Search Performance of SHADE Using Linear Population Size Reduction. In Proceedings of the 2014 IEEE Congress on Evolutionary Computation (CEC), Beijing, China, 6–11 July 2014.

39. Pedrosa Silva, R.; Lopes, R.; Guimarães, F. Self-adaptive mutation in the Differential Evolution: Self- * search. In Proceedings of the Genetic and Evolutionary Computation Conference, GECCO'11, Dublin, Ireland, 12–16 July 2011; pp. 1939–1946.

40. Ilonen, J.; Kamarainen, J.K.; Lampinen, J. Differential evolution training algorithm for feedforward neural networks. *Neural Process. Lett.* **2003**, *17*, 93–105. [CrossRef]

41. Masters, T.; Land, W. A new training algorithm for the general regression neural network. In Proceedings of the 1997 IEEE International Conference on Systems, Man, and Cybernetics, Orlando, FL, USA, 12–15 October 1997; Volume 3, pp. 1990–1994.

42. Schraudolph, N.N.; Belew, R.K. Dynamic parameter encoding for genetic algorithms. *Mach. Learn.* **1992**, *9*, 9–21. [CrossRef]

43. Mattiussi, C.; Dürr, P.; Floreano, D. Center of Mass Encoding: A Self-adaptive Representation with Adjustable Redundancy for Real-valued Parameters. In Proceedings of the Genetic and Evolutionary Computation Conference, London, UK, 7–11 July 2007; ACM: New York, NY, USA, 2007; pp. 1304–1311.

44. Mancini, L.; Milani, A.; Poggioni, V.; Chiancone, A. Self regulating mechanisms for network immunization *AI Commun.* **2016**, *29*, 301–317. [CrossRef]

45. Franzoni, V.; Chiancone, A. A Multistrain Bacterial Diffusion Model for Link Prediction *Int. J. Pattern Recognit. Artif. Intell.* **2017**, *31*. [CrossRef]

46. Heidrich-Meisner, V.; Igel, C. Neuroevolution strategies for episodic reinforcement learning. *J. Algorithms* **2009**, *64*, 152–168. [CrossRef]

47. Leema, N.; Nehemiah, H.K.; Kannan, A. Neural network classifier optimization using Differential Evolution with Global Information and Back Propagation algorithm for clinical datasets. *Appl. Soft Comput.* **2016**, *49*, 834–844. [CrossRef]

48. Quade, D. Using weighted rankings in the analysis of complete blocks with additive block effects. *J. Am. Stat. Assoc.* **1979**, *74*, 680–683. [CrossRef]

Article

Dynamic Parallel Mining Algorithm of Association Rules Based on Interval Concept Lattice

Yafeng Yang [1,2,3], Ru Zhang [4] and Baoxiang Liu [1,2,]*

1   College of Science, North China University of Science and Technology, 21 Bohai Road,
    Tangshan 063210, China
2   Hebei Key Laboratory of Data Science and Application, 21 Bohai Road, Tangshan 063210, China
3   Tangshan Key Laboratory of Engineering Computing, 21 Bohai Road, Tangshan 063210, China
4   Department of mathematics and information sciences, Tangshan Normal University, No. 156 Jianshe North
    Road, Tangshan 063009, China
*   Correspondence: www1673@163.com

Received: 19 May 2019; Accepted: 17 July 2019; Published: 19 July 2019

Abstract: An interval concept lattice is an expansion form of a classical concept lattice and a rough concept lattice. It is a conceptual hierarchy consisting of a set of objects with a certain number or proportion of intent attributes. Interval concept lattices refine the proportion of intent containing extent to get a certain degree of object set, and then mine association rules, so as to achieve minimal cost and maximal return. Faced with massive data, the structure of an interval concept lattice is more complex. Even if the lattice structures have been united first, the time complexity of mining interval association rules is higher. In this paper, the principle of mining association rules with parameters is studied, and the principle of a vertical union algorithm of interval association rules is proposed. On this basis, a dynamic mining algorithm of interval association rules is designed to achieve rule aggregation and maintain the diversity of interval association rules. Finally, the rationality and efficiency of the algorithm are verified by a case study.

Keywords: interval concept lattice; association rules; mining algorithm; vertical union

---

1. Introduction

A concept lattice [1] is a conceptual hierarchy constructed according to the binary relationship between objects and attributes in data sets. As an effective tool for knowledge representation, a concept lattice is widely used in knowledge discovery, rule mining, information retrieval, and other fields because of its accuracy and completeness [2–4].

Concept lattice theory mainly focuses on the following aspects: A concept lattice extent model [5–7], concept lattice construction and rule extraction [8–10], concept lattice merging [11–15], concept lattice reduction [16,17], concept lattice modification [18], etc.

In a classic concept lattice, the concept extents have all the attributes or only one attribute, sometimes. Hence, the support and confidence degree of the extracted association rules would be reduced greatly. To solve this problem, the authors have put forward a new concept lattice structure: An interval concept lattice, and the construction methods, compression, and maintenance of lattice structure were studied [19–22].

From the perspective of a concept lattice, the relationship between intents is association rules, while the relationship between extents is its embodiment. A concept lattice is the unity of intent and extent, and the relationship between its nodes also reflects the relationship between the generalization and specialization of concepts. Therefore, a concept lattice is suitable for the application of a basic data structure in association rules mining. Many scholars have conducted in-depth research on rule mining based on a concept lattice [23–28].

Previously, we studied the structure characteristics of an interval concept lattice and gave two measurement standards of the uncertainty rule—precision and uncertainty. Then, a mining model of interval association rule with parameters was constructed [29]. The algorithm can mine and optimize rules according to the adjustment of parameters, which is of great significance to the mining of rules with uncertainties. Next, the complex relationship between interval parameters and association rules and the optimization algorithm of the rule base were given [30]. By adjusting the parameters, the purpose of controlling and optimizing rules was achieved.

In the era of data explosion, people's demand for data processing is getting higher and higher. The real-time updating of data requires the efficient processing of dynamic data. For example, in the supermarket shopping system, the massive transaction information generated every day can only mine local association rules, but cannot provide a timely and accurate decision-making plan for decision-makers as a whole. However, the time and space complexity of the process will increase rapidly with the increase of the amount of data, and the mining association rules will be missing. Therefore, it is necessary to study the dynamic mining of interval association rules in order to grasp uncertain rules in real time.

The authors have studied the consistency of interval concept lattices, discussed the decision theorem of the concept of consistent intent, and designed a vertical union algorithm of interval concept lattices based on the breadth-first principle [31]. Furthermore, the sequential traversal method was used to scan the lattice structure, and a union algorithm of interval concept lattices was proposed from a transverse point of view [32]. Based on the research results of the union algorithm of interval concept lattice, this paper carries out the dynamic mining of interval association rules.

2. Concepts and Methods

2.1. Interval Concept Lattice

Definition 1 ([33]). *For the formal context* (U, A, R), *where* $U = \{x_1, x_2, \cdots, x_3\}$ *is the object sets and each* $x_i (i \leq n)$ *denotes an object;* $A = \{a_1, a_2, \cdots, a_m\}$ *is the attribute set, and each* $a_j (j \leq m)$ *denotes an attribute;* R *is the binary relationship between* U *and* A. $R \subseteq U \times a$. *If* $(x, a) \in R$, *then we record that* x *has the attribute* a, *and write as* xRa.

Definition 2 ([33]). *For the formal context* (U, A, R), *operators* f, g *are defined as follows:*

$\forall x \in U, f(x) = \{y | \forall y \in A, xRy\}$, *i.e.,* f *is the mapping between* x *and its attributes;*
$\forall y \in a, g(y) = \{x | \forall x \in U, xRy\}$, *i.e.,* g *is the mapping between* y *and its objects.*

Definition 3 ([33]). *For the formal context* (U, A, R), *if* $f(X) = Y, g(Y) = X$ *for* $X \subseteq U, Y \subseteq A$, *then the sequence* $< X, Y >$ *is called a formal concept, or concept for short.* X *is the extent and* Y *is the intent.*

Rough concept lattice $RL(U, A, R)$ based on rough set theory was studied in Reference [7], where the upper approximation extent and lower approximation extent refer to the maximal concept set and the minimal concept set respectively which have all the attributes in $Y \subseteq A$.

Definition 4 ([33]). *For the formal context* (U, A, R) *and its rough concept lattice* $RL(U, A, R)$, (M, N, Y) *is the rough concept. Set an interval* $[\alpha, \beta]$ $(0 \leq \alpha \leq \beta \leq 1)$, *then* α *upper bound extent* M^{α} *and* β *lower bound extent* M^{β} *are:*

$$M^{\alpha} = \left\{ x \middle| x \in M, \left| f(x) \cap Y \right| / |Y| \geq \alpha, 0 \leq \alpha \leq 1 \right\} \tag{1}$$

$$M^{\beta} = \left\{ x \middle| x \in M, \left| f(x) \cap Y \right| / |Y| \geq \beta, 0 \leq \alpha \leq \beta \leq 1 \right\} \tag{2}$$

Y is the concept intent and |Y| is the number of elements in Y, that is base number. M^α refers to the objects which may be covered by $\alpha \times |Y|$ attributes or more in Y. M^β refers to the objects which may be covered by $\beta \times |Y|$ attributes or more in Y.

Definition 5 ([33]). *Let (U, A, R) be a formal context and (M^α, M^β, Y) be an interval concept. Then, Y is the intent; M^α is the α upper bound extent and M^β is the β lower bound extent.*

Definition 6. *Suppose that (U, A, R) has two interval concepts, $(M_1^\alpha, M_1^\beta, Y_1)$ and $(M_2^\alpha, M_2^\beta, Y_2)$. If the two meet $Y_1 \subseteq Y_2$, $|Y_2| - |Y_1| = 1$, $M_1^\alpha = M_2^\alpha$ and $M_1^\beta = M_2^\beta$, then $(M_1^\alpha, M_1^\beta, Y_1)$ is called the redundant concept.*

Definition 7. *Suppose that (U, A, R) has an interval concept, (M^α, M^β, Y). If it meets $M^\alpha = M^\beta = \varnothing$, then (M^α, M^β, Y) is called the empty concept.*

Definition 8. *Suppose (U, A, R) has an interval concept, $C = (M^\alpha, M^\beta, Y)$. If C is neither the redundant concept nor the empty concept, then C is called the existence concept. $L_\alpha^\beta(U, A, R)$ is a collection of all the existence concepts.*

Definition 9. $\overline{L_\alpha^\beta}(U, A, R)$ *refers to all the $[\alpha, \beta]$ interval concepts, which include: Existence concepts, redundant concepts, and empty concepts, that is:*

$$(M_1{}^\alpha, M_1{}^\beta, Y_1) \leq (M_2{}^\alpha, M_2{}^\beta, Y_2) \Leftrightarrow Y_1 \supseteq Y_2, \tag{3}$$

Then "$\leq$"is called the partial order relationship of $\overline{L_\alpha^\beta}(U, A, R)$.

Definition 10. *If all the concepts in $\overline{L_\alpha^\beta}(U, A, R)$ meet "$\leq$", then $L_\alpha^\beta(U, A, R)$ is called interval concept lattice on the formal context (U, A, R).*

Definition 11. *In the interval concept lattice $L_\alpha^\beta(U, A, R)$, if $C = (M^\alpha, M^\beta, Y) \in L_\alpha^\beta(U, A, R)$, then the layer of the Lattice Structure is $|A| + 1$ and node C is at Layer $|Y|$. In particular, when $Y = \varnothing$, C was recorded on the zeroth layer.*

2.2. Interval Association Rules

Formal context (U, A, R) can describe a database, where U represents an object set; A represents an attribute set. For $x \in U$, $a \in A$, xRa represents the item-set where a belongs to x.

Definition 12. *Given the minimal support threshold θ, for any interval concept node C, if the number of objects in the upper bound extent is not less than $|U| \times \theta$, then C is called the α-upper bound frequent node, and the corresponding Y is called the α-upper bound frequent item-set; if the number of objects in the lower bound extent is not less than $|U| \times \theta$, then C is called the β-lower bound frequent node, the corresponding Y is called the β-lower bound frequent item-set.*

The father and son concept in the interval concept lattice does not have a specific relationship in frequency, which is different from the classical concept lattice.

If the association rule $A \Rightarrow B$ corresponds to the interval concept node $(C1, C2)(C1 = (M_1^\alpha, M_1^\beta, Y_1)$, $C2 = (M_2^\alpha, M_2^\beta, Y_2))$ and $C1 \geq C2$, then rule $A \Rightarrow B$ is generated by node binary $(C1, C2)$.

The α-upper bound association rules and the β-lower bound association rules can be extracted by two lower bound extents of the interval concept. The calculation methods of confidence and support are as follows:

The α-upper bound rule $A \Rightarrow B$:

$$Conf(A \Rightarrow B) = \left|M_2^{\alpha}\right| / \left|M_1^{\alpha}\right| \tag{4}$$

$$Support(A \Rightarrow B) = \left|M_2^{\alpha}\right| / |U| \tag{5}$$

The β-lower bound $A \Rightarrow B$:

$$Conf(A \Rightarrow B) = \left|M_2^{\beta}\right| / \left|M_1^{\beta}\right| \tag{6}$$

$$Support(A \Rightarrow B) = \left|M_2^{\beta}\right| / |U| \tag{7}$$

Definition 13. *Given the minimal support threshold θ and the minimal confidence threshold c. Node binary $(C1, C2)$ is called α-upper bound candidate binary which consists of two frequent concept nodes $C1 = (M_1^{\alpha}, M_1^{\beta}, Y_1)$ and $C2 = (M_2^{\alpha}, M_2^{\beta}, Y_2)$, where $M_2^{\alpha} \subseteq M_1^{\alpha}$ and $\left|M_2^{\alpha}\right| / \left|M_1^{\alpha}\right| \geq c$. When $(C1, C2)$ meets $M_2^{\beta} \subseteq M_1^{\beta}$ and $\left|M_2^{\beta}\right| / \left|M_1^{\beta}\right| \geq c$, it is called β-lower bound candidate binary.*

Definition 14. *For interval association rule $A \Rightarrow B$, if $A \cup B$ is frequent item-sets, $Support(A \Rightarrow B) \geq \theta$ and $Conf(A \Rightarrow B) \geq c$, i.e., $\left|P(A \cup B)\right| / \left|P(A)\right| \geq c$, then it is called the strong association rule.*

Definition 15. *If $A \Rightarrow B$ is the strong association rule, then $C \Rightarrow D$ must be the strong association rule, then we say "$A \Rightarrow B$ can derive $C \Rightarrow D$".*

Theorem 1. *If $C \subset D$, then $A \Rightarrow B$ can derive $A \Rightarrow C$.*

Proof. $\quad C \subset B \Rightarrow |C| < |B| \Rightarrow |A \cup C| < |A \cup B| \Rightarrow Support(A \Rightarrow C) = |C| < |U| \quad < Support(A \Rightarrow B) = |B| / |U|$ and $Conf(A \Rightarrow C) = |A \cup C| / |A| < Conf(A \Rightarrow B) = |A \cup B| / |A|$. $\square$

Theorem 2. *In the interval concept lattice, if $(C1, C2)$ and $(C1, C3)$ are candidate binaries and $C3 > C2$, then all the rules in $Rules(C1, C3)$ can be derived from $Rules(C1, C2)$.*

Definition 16. *Suppose $A \Rightarrow B$ is α-upper bound association rule derived from $(C1, C2)$. The upper bound extent of $C1$ is $M_1^{\alpha} = \{x_1, x_2, \ldots, x_m,\}$. The intent of $C1$ is Y_1. The upper bound extent of $C2$ is $M_2^{\alpha} = \{o_1, o_2, \ldots, o_m,\}$. The intent of $C2$ is Y_2. The precision of $A \Rightarrow B$ is*

$$PD_{A \Rightarrow B} = \min\left\{ \min_{i=1}^{m} \frac{|x_i.Y \cap Y_1|}{|Y_1|}, \min_{i=1}^{n} \frac{|o_i.Y \cap Y_2|}{|Y_2|} \right\} \tag{8}$$

The uncertainty of $A \Rightarrow B$ is $UD_{A \Rightarrow B} = 1 - PD_{A \Rightarrow B}$.

Definition 17. *Let $\Omega = \{Rule1, Rule2, \cdots, Rulek\}$ denotes to α-rules set, the uncertainty of $Rulei$ is $UD_{\alpha-Ri}$, then the uncertainty of α-rules sets is*

$$UD_{\alpha-Rluesset} = \max_{i=1}^{k}(UD_{\alpha-Ri}) \tag{9}$$

Let $\Omega = \{Rule1, Rule2, \cdots, Rulem\}$ denote to β-rules set, the uncertainty of $Rulej$ is $UD_{\beta-Ri}$, then the uncertainty of β-rules sets is

$$UD_{\beta-Rluesset} = \max_{j=1}^{m}(UD_{\beta-Ri}) \tag{10}$$

and the uncertainty of interval association rules is $UD = \max(UD_{\alpha-Rluesset}, UD_{\beta-Rluesset})$.

Theorem 3. *Suppose the minimal support threshold θ and the minimal confidence threshold c. In the same context, when one of the interval parameters α and β is unchanged and the other is larger, the process of extracting association rules will change as follows:*

(1)   *The number of frequent nodes generated does not increase;*
(2)   *The node with the largest intent in frequent nodes does not increase in intent cardinality.*
(3)   *The number of candidate binary arrays generated does not increase;*
(4)   *The number of generated association rules does not increase.*

3. Algorithm and Results

In the mining method of interval association rules, the lattice structure is united with the vertical union algorithm first, and then the association rules are extracted by using the parametric association rules mining algorithm. Because the structure of the interval concept lattice is complex, the memory space required is large, and the time complexity of mining interval association rules after vertical merging is high, this method does not meet the requirements of the current era of large data. In this section, the principle of a vertical union algorithm of interval association rules is proposed, and on this basis, a dynamic mining algorithm of interval association rules is designed.

3.1. Vertical Union Principle of Interval Association Rules

If the interval concept lattices $\overline{L_\alpha^\beta(U_1, A_1, R_1)}$ and $\overline{L_\alpha^\beta(U_2, A_2, R_2)}$ are consistent and $A_1 = A_2 = A$. $U_1 \cap U_2 = \phi$, then $\overline{L_\alpha^\beta(U, A, R)}$ can be gotten through the union of the two. Suppose the minimal support threshold θ and the minimal confidence threshold c, we can extract the association rules of $\overline{L_\alpha^\beta(U_1, A, R_1)}$ and $\overline{L_\alpha^\beta(U_2, A, R_2)}$. Here, $\alpha - Rluesset^*$ is the upper bound association rules set derived from $\overline{L_\alpha^\beta(U_1, A, R_1)}$. $\alpha - Rluesset^{**}$ is the upper bound association rules set derived from $\overline{L_\alpha^\beta(U_2, A, R_2)}$; $\alpha - Rluesset$ is the upper bound association rules set derived from $\overline{L_\alpha^\beta(U, A, R)}$.

$$C_1^* = (M_1^{\alpha*}, M_1^{\beta*}, Y_1),\ C_2^* = (M_2^{\alpha*}, M_2^{\beta*}, Y_2)\ \text{and}\ C_1^*, C_2^* \in \overline{L_\alpha^\beta(U_1, A, R_1)};$$

$$C_1^{**} = (M_1^{\alpha**}, M_1^{\beta**}, Y_1),\ C_2^{**} = (M_2^{\alpha**}, M_2^{\beta**}, Y_2)\ \text{and}\ C_1^{**}, C_2^{**} \in \overline{L_\alpha^\beta(U_2, A, R_2)};$$

$$C_1 = (M_1^\alpha, M_1^\beta, Y_1),\ C_2 = (M_2^\alpha, M_2^\beta, Y_2)\ \text{and}\ C_1, C_2 \in \overline{L_\alpha^\beta(U, A, R)}.$$

Now, taking the upper bound extent as an example (the union principle of lower bound extent and upper bound extent), the vertical union of interval association rules $\alpha - Rluesset^*$ and $\alpha - Rluesset^{**}$ is carried out as follows:

Theorem 4. *If C_1^*, C_2^* constitutes the association rule, and $Rules(C_1^*, C_2^*) \in \alpha - Rluesset^*$; C_1^{**}, C_2^{**} the constitutes association rule, and $Rules(C_1^{**}, C_2^{**}) \in \alpha - Rluesset^{**}$, then $Rules(C_1, C_2) \in \alpha - Rluesset$.*

Proof. $Rules(C_1^*, C_2^*) \in \alpha - Rluesset^*$ and $Rules(C_1^{**}, C_2^{**}) \in \alpha - Rluesset^{**}$, so:
Relation 1: $M_2^{\alpha*} \subseteq M_1^{\alpha*}$ and $M_2^{\alpha**} \subseteq M_1^{\alpha**}$, then $M_2^{\alpha*} \cup M_2^{\alpha**} \subseteq M_1^{\alpha*} \cup M_1^{\alpha**}$, i.e., $M_2^\alpha \subseteq M_1^\alpha$.

Relation 2: $\dfrac{|M_2^{\alpha*}|}{|M_1^{\alpha*}|} \geq c$ and $\dfrac{|M_2^{\alpha**}|}{|M_1^{\alpha**}|} \geq c$, then

$$|M_2^{\alpha*}| + |M_2^{\alpha**}| \geq c(|M_1^{\alpha*}| + |M_1^{\alpha**}|),\ \frac{|M_2^{\alpha*}| + |M_2^{\alpha**}|}{|M_1^{\alpha*}| + |M_1^{\alpha**}|} \geq c, \text{i.e.,}\ \frac{|M_2^\alpha|}{|M_1^\alpha|} \geq c.$$

Then $Rules(C_1, C_2) \in \alpha - Rluesset$ can be derived from two relations. $\square$

Theorem 5. *If at least one of the two association rules* $Rules(C_1^*, C_2^*)$ *and* $Rules(C_1^{**}, C_2^{**})$ *in* $\overline{L_\alpha^\beta(U_1, A, R_1)}$ *and* $\overline{L_\alpha^\beta(U_2, A, R_2)}$ *does not exist, then there must be no* $Rules(C_1, C_2)$ *in* $\alpha - Rluesset.$

Because the vertical union of interval association rules is based on the objects set with certain proportion attributes, it is necessary to introduce support, confidence, accuracy, and uncertainty to measure accurately.

The upper bound frequencies of C_1^*, C_2^* are θ_1^* and θ_2^*, respectively.

The upper bound frequencies of C_1^{**}, C_2^{**} are θ_1^{**} and θ_2^{**}, respectively.

The upper bound frequencies of C_1, C_2 are θ_1 and θ_2, respectively.

The support degree, confidence degree, accuracy, and uncertainty of $Rules(C_1, C_2)$ obtained by vertical union of association rules $Rules(C_1^*, C_2^*)$ and $Rules(C_1^{**}, C_2^{**})$ can be deduced from frequency to its solution formula, as follows.

Theorem 6. *The support degree of* $Rules(C_1, C_2)$ *is:*

$$Support(C_1 \Rightarrow C_2) = \frac{|U_1|\theta_2^* + |U_2|\theta_2^{**}}{|U_1| + |U_2|} \tag{11}$$

Theorem 7. *The confidence degree of* $Rules(C_1, C_2)$ *is*

$$Conf(C_1 \Rightarrow C_2) = \frac{|U_1|\theta_2^* + |U_2|\theta_2^{**}}{|U_1|\theta_1^* + |U_2|\theta_1^{**}} \tag{12}$$

Theorem 8. $Rules(C_1, C_2) = Min\{Rules(C_1^*, C_2^*), Rules(C_1^{**}, C_2^{**})\}.$

3.2. Dynamic Mining Algorithms for Interval Association Rules

3.2.1. Algorithm Design

According to Theorems 4 and 5, the result of the vertical union of interval association rules only occurs in the data with the same interval association rules and adjacent data. In order to ensure that the mined interval association rules are not lost, the algorithm retains the non-united rules on the basis of a vertical union of the same rules, which facilitates the mining of interval association rules later, and measures the occurrence frequency and frequency of the same rules with frequency. The basic idea of the algorithm is that whenever an interval association rule is generated, it is transformed into an array representation, and it is mined with the set of interval rules that have been united and retain the non-united rules, so that the interval rules can be aggregated again and again.

In order to distinguish different rules, association rules mined from interval concept lattices are stored in the form of arrays $RS[i] = \{Rule, FN_1, FN_2, U, Support, Conf, PD, UD, Flag, Num\}$.

$RS[i]$ represents the ith association rule in the interval association rules set RS;

$Rule$ represents the interval association rules $Rule(C_x, C_y)$;

FN_1, FN_2 represents the frequency degree of frequency nodes C_x and C_y in $Rule(C_x, C_y)$;

U represents the number of objects in the context;

$Support, Conf, PD$ and UD represents the support, confidence, accuracy, and uncertainty degree of $Rule(C_x, C_y)$;

$Flag$ marked according to whether rules have been merged or not. $Flag = 0$ denotes that rules are not united vertically; $Flag = 1$ indicates that the rules have been vertically united; Num denotes the number of occurrences of rule $Rule(C_x, C_y)$ in previously united rule sets.

Based on the above algorithm principle and analysis, we design a dynamic extraction and union algorithm of interval association rules, DMA (Dynamic Mining Algorithm). See Algorithm 1.

Algorithm 1. DMA (Dynamic Mining Algorithm)

Input: Association rule sets $RS_1, RS_2, \ldots RS_k \ldots$

Output: Association rule set RS

Step1 $RS = RS_1$;

Step2 The interval association rules in the rule set RS_k is stored in the form of arrays. Set RS^* and initialize. Comparing the *Rule* of rule set RS and RS_k, we unit interval association rules vertically according to Theorems 4 and 5. For the rules that have been united in RS and RS_k, let $Flag = 1$ and put the united rule number in RS^*. According to Theorems 6–8, calculate the frequency,*Support, Conf, PD,* and *UD* of C_x and C_y in RS^*. *Flag*=1, $Num = Num + 1$. Delete the rules of $Flag = 1$ in RS and RS_k, and renumbering. Let rules number in RS^* add the renumbering number in RS. Putting the remain rules of RS into RS^*; then renumbering in RS_k, and putting the remain rules of RS_k into RS^*.

Rule Vertical Union (RS, RS_k)

1   {

2     $g = 0$;

3     $RS[g]^* = \{Rule, FN_1, FN_2, U, Support, Conf, \quad PD, UD, Flag, Num\}$

4     $\{ Rule = \varnothing$;

5     $FN_1 = FN_2 = 0$;

6     $U = RS[1].U + RS_k[1].U$;

7     $Support = Conf = PD = UD = Flag = 0$;

8     $Num = 1$;

9   }

10   For (i=1;$|RS|$; i++)

11   {For (j=1;$|RS_k|$; j++)

12   If($RS[i].Flag = RS_k[j].Flag = 0\|$ $RS[i].Rule = RS_k[j].Rule$)

13      $\{ g + 1$;

14       $Num + 1$;

15       $RS[i].Flag = 1$;

16       $RS_k[j].Flag = 1$;

17       $RS[g]^*.FN_1 = \frac{RS[i].U*RS[i].FN_1 + RS_k[j].U*RS_k[j].FN_1}{RS[i].U + RS_k[j].U}$

18       $RS[g]^*.FN_2 = \frac{RS[i].U*RS[i].FN_2 + RS_k[j].U*RS_k[j].FN_2}{RS[i].U + RS_k[j].U}$

19    $RS[g]^*.Support = RS[g]^*.FN_2$

20    $RS[g]^*.Conf = \frac{RS[i].U*RS[i].FN_2 + RS_k[j].U*RS_k[j].FN_2}{RS[i].U*RS[i].FN_1 + RS_k[j].U*RS_k[j].FN_1}$

21    $RS[g]^*.PD = \min\{RS[i].PD, RS_k[j].PD\}$;

22     $RS[g]^*.UD = 1 - RS[g]^*.PD$;

23   }

24   }

25   For each $RS[i]$ in RS;

26   {If $RS[i].Flag = 0$;

27     $g + 1$;

28     $RS^*[g] = RS[i]$;}

29   For each $RS_k[j]$ in RS_k;

30   {If $RS_k[j].Flag = 0$;

31     $g + 1$;

32    $RS^*[g] = RS_k[j]$;}

33   }

Step3 $RS = RS^*$

3.2.2. Algorithm Analysis

The main content of DMA is embodied in the function Rule Vertical Union (RS, RS_k). Lines 1 to 9 implement the initialization of parallel rule set RS^*. Lines 10–12 nested for loop implements searching for interval association rules in RS and RS_k that correspond to the same rules and are not united. Lines 13 to 16 number the interval association rules found in RS^*. Num+1 is used to count the rule, and the corresponding rule *Flag* in united RS, RS_k is recorded as 1. Lines 17 to 25 implement the assignment of interval association rule $RS^*[g]$. In lines 26–34, the unconsolidated rules in RS, RS_k are put into RS^* to maintain the diversity of the united rule set.

Compared with Step 2, the algorithm achieves the step requirements, and the assignment of Step 1 and Step 3 make the DMA algorithm dynamic, so that the algorithm has integrity and correctness. Compared with the general method, the algorithm realizes the merging of rules to rules, eliminating the process of vertical merging of interval concept lattices, thus greatly reducing the time complexity and space complexity of the implementation process.

If the number of association rules in two sub lattices is n and m respectively, the time complexity of the algorithm is less than $O(n \times m) + O(n) + O(m)$. Compared with the interval association rule mining algorithm with parameters, the algorithm is more efficient.

3.3. Example Study

Set the formal contexts as shown in Tables 1 and 2:

Table 1. Formal context FC_1.

	a	b	c	d	e
1	1	1	1	0	0
2	0	0	0	1	0
3	1	1	0	1	0
4	1	0	1	0	1

Table 2. Formal context FC_2.

	a	b	c	d	e
(1)	1	0	1	1	0
(2)	1	1	0	1	0
(3)	0	1	0	1	1

Set $\alpha = 0.6, \beta = 0.7, \theta = 50\%, c = 60\%$.

The DMA algorithm is used to mine interval association rules in parallel. Take the upper bound interval association rules as an example (the mining of lower bound association rules is similar).

Among them, the upper bound frequent nodes from FC_1 and FC_2 are shown in Tables 3 and 4 respectively, and the corresponding upper bound association rules are shown in Tables 5 and 6 respectively.

Table 3. The frequent nodes of FC_1.

Frequent Node	Frequent Degree	Frequent Node	Frequent Degree
a	75%	acd	100%
b	50%	ace	50%
c	75%	ade	50%
d	50%	bcd	75%
ab	50%	bce	50%
ac	50%	cde	50%
abc	75%	abcd	50%
abd	50%	abce	50%
abe	75%	abcde	75%

Table 4. The frequent nodes of FC_2.

Frequent Node	Frequent Degree	Frequent Node	Frequent Degree
a	67%	acd	67%
b	67%	ade	100%
d	100%	bcd	100%
ad	67%	bde	67%
bd	67%	cde	67%
abc	67%	abcd	67%
abd	100%	abde	67%
abe	67%	abcde	100%

Table 5. The previous rules and its measurement results on FC_1.

Rule	Support	Confidence	Accuracy	Uncertainty	Frequent Number
$a \Rightarrow bcde$	75%	100%	60%	40%	1
$abc \Rightarrow de$	75%	100%	60%	40%	1
$abe \Rightarrow cd$	75%	100%	60%	40%	1
$acd \Rightarrow be$	75%	75%	60%	40%	1
$a \Rightarrow bce$	50%	67%	75%	25%	1
$c \Rightarrow abe$	50%	67%	75%	25%	1
$ac \Rightarrow be$	50%	100%	75%	25%	1
$abc \Rightarrow e$	50%	67%	67%	33%	1
$abe \Rightarrow c$	50%	67%	67%	33%	1
$ace \Rightarrow b$	50%	100%	67%	33%	1
$a \Rightarrow bcd$	50%	67%	75%	25%	1
$b \Rightarrow acd$	50%	100%	75%	25%	1
$ab \Rightarrow cd$	50%	100%	75%	25%	1
$abc \Rightarrow d$	50%	67%	67%	33%	1
$abd \Rightarrow c$	50%	100%	67%	33%	1
$bcd \Rightarrow a$	50%	67%	67%	33%	1
$c \Rightarrow de$	50%	67%	67%	33%	1
$c \Rightarrow be$	50%	67%	67%	33%	1
$a \Rightarrow de$	50%	67%	67%	33%	1
$a \Rightarrow ce$	50%	67%	67%	33%	1
$c \Rightarrow ae$	50%	67%	67%	33%	1
$ac \Rightarrow e$	50%	100%	67%	33%	1
$a \Rightarrow be$	75%	100%	67%	33%	1
$a \Rightarrow bd$	50%	67%	67%	33%	1
$b \Rightarrow ae$	50%	100%	67%	33%	1
$ab \Rightarrow e$	50%	100%	67%	33%	1
$a \Rightarrow bc$	75%	100%	67%	33%	1
$a \Rightarrow c$	50%	67%	100%	0%	1
$c \Rightarrow a$	50%	67%	100%	0%	1
$a \Rightarrow b$	50%	67%	100%	0%	1
$b \Rightarrow a$	50%	100%	100%	0%	1

The DMA algorithm is used to mine the previous association rules of FC_1 and FC_2 in parallel. The parallel results of the different association rules are the same part of the combined association rules deleted in Tables 5 and 6, respectively.

The parallel results of the same association rules are shown in Table 7.

Table 6. The previous rules and its measurement results on FC_2.

Rule	Support	Confidence	Accuracy	Uncertainty	Frequent Number
$d \Rightarrow abce$	100%	100%	60%	40%	1
$abd \Rightarrow ce$	100%	100%	60%	40%	1
$ade \Rightarrow bc$	100%	100%	60%	40%	1
$bcd \Rightarrow ae$	100%	100%	60%	40%	1
$b \Rightarrow ade$	67%	100%	75%	25%	1
$d \Rightarrow abe$	67%	67%	75%	25%	1
$bd \Rightarrow ae$	67%	100%	75%	25%	1
$abd \Rightarrow e$	67%	67%	67%	33%	1
$abe \Rightarrow d$	67%	100%	67%	33%	1
$ade \Rightarrow b$	67%	67%	67%	33%	1
$bde \Rightarrow a$	67%	100%	67%	33%	1
$a \Rightarrow bcd$	67%	100%	75%	25%	1
$d \Rightarrow abc$	67%	67%	75%	25%	1
$ad \Rightarrow bc$	67%	100%	75%	25%	1
$abc \Rightarrow d$	67%	100%	67%	33%	1
$abd \Rightarrow c$	67%	67%	67%	33%	1
$acd \Rightarrow b$	67%	100%	67%	33%	1
$bcd \Rightarrow a$	67%	67%	67%	33%	1
$d \Rightarrow ce$	50%	67%	67%	33%	1
$b \Rightarrow de$	67%	100%	67%	33%	1
$d \Rightarrow be$	67%	67%	67%	33%	1
$bd \Rightarrow e$	67%	100%	67%	33%	1
$d \Rightarrow bc$	100%	100%	67%	33%	1
$d \Rightarrow ae$	100%	100%	67%	33%	1
$a \Rightarrow cd$	67%	100%	67%	33%	1
$d \Rightarrow ac$	67%	67%	67%	33%	1
$ad \Rightarrow c$	67%	100%	67%	33%	1
$b \Rightarrow ae$	67%	100%	67%	33%	1
$d \Rightarrow ab$	100%	100%	67%	33%	1
$a \Rightarrow bc$	67%	100%	67%	33%	1

Table 7. The previous rules set and its measurement results after vertical union.

Rule	Support	Confidence	Accuracy	Uncertainty	Frequent Number
$a \Rightarrow bcd$	57%	80%	75%	25%	2
$abc \Rightarrow d$	57%	80%	67%	33%	2
$abd \Rightarrow c$	57%	80%	67%	33%	2
$bcd \Rightarrow a$	57%	67%	67%	33%	2
$a \Rightarrow bc$	71%	100%	67%	33%	2

4. Discussion

Considering the characteristics of interval concept lattices, the vertical union method of interval concept lattices and the principle of mining association rules with parameters were studied in this paper. Considering the intrinsic relationship between interval association rules, this paper presents some metrics, such as the confidence and credibility of united rules, and proposes a dynamic mining algorithm for interval association rules, which realizes rule aggregation and keeps the diversity of interval association rules. The rationality and efficiency of the algorithm are proved by algorithm analysis and case study, which provides timely, effective and abundant decision information for decision makers in data analysis. The next step is to optimize the algorithm for incomplete information systems in the context of large data. In the mining process, the stability and timing choice of the rule union is another problem to be studied in depth.

Author Contributions: Y.Y. contributed to the main body of the paper. The main models and algorithms were constructed; R.Z. was mainly responsible for the writing of the paper, colleagues performed data processing; B.L. guided the overall process of the paper.

Funding: This research was funded by the National Natural Science Foundation of China (61370168, 61472340), Natural Science Foundation of Hebei Province (F2016209344).

Acknowledgments: The authors thank the support of Hebei Key Laboratory for Data Science and Application and the technical assistance of Chunying Zhang and Lihong Li from North China University of Science and Technology.

Conflicts of Interest: The authors declare no conflicts of interest.

References

1. Wille, R. Restructing lattice theory: An approach based on hierarchies of concepts. In *Ordered Sets*; Rival, I., Ed.; Reidel: Dordrecht, The Netherlands, 1982; pp. 445–470.
2. Jiang, C. Applications of concept lattice in the digital library. *Inf. Sci.* **2010**, *28*, 1908–1911.
3. Cole, R.; Stumme, G. CEM—A conceptual email manager. In Proceedings of the 7th International Conference on Conceptual Structures (ICCS'2000), Darmstadt, Germany, 14–18 August 2000; Springer: Berlin/Heidelberg, Germany, 2000.
4. Ganter, B.; Wille, R. *Formal Concept Analysis: Mathematical Foundations*; Springer: Berlin, Germany, 1999; pp. 53–85.
5. Zhao, W. *Research of Data Mining Based on Concept Lattice and Its Extended Models*; HeFei University of Technology: Hefei, China, 2002.
6. Zhang, J.; Zhang, S.; Zheng, L. Weighted concept lattice and incremental construction. *Pattern Recognit. Artif. Intell.* **2005**, *18*, 171–176.
7. Yang, H.; Zhang, J. Rough concept lattice and construction arithmetic. *Comput. Eng. Appl.* **2007**, *43*, 172–175.
8. Ho, T.B. An approach to concept formation based on formal concept analysis. *IEICE Trans. Inf. Syst.* **1995**, *78*, 553–559.
9. Liu, B.; Li, Y. Construction principles and algorithms of concept lattice generated by random decision formal context. *Comput. Sci.* **2013**, *40*, 90–92.
10. Godin, R. Incremental concept formation algorithm based on Galois (concept) lattices. *Comput. Intell.* **1995**, *11*, 246–267. [CrossRef]
11. Li, Y.; Liu, Z.; Chen, L.; Xu, X.; Cheng, W. Horizontal union algorithm of multiple concept lattices. *Acta Electron. Sin.* **2004**, *32*, 1849–1854.
12. Zhi, H.; Zhi, D.; Liu, Z. Theory and algorithm of concept lattice union. *Acta Electron. Sin.* **2010**, *38*, 455–459.
13. Zhang, L.; Shen, X.; Han, D.; An, G. Vertical union algorithm of concept lattice based on synonymous concept. *Comput. Eng. Appl.* **2007**, *43*, 95–98.
14. Xu, M.; He, Z.; Liu, Y.; Li, J.; Zhu, Q. Bidirectional incremental formation algorithm of concept lattice. *J. Chin. Comput. Syst.* **2014**, *1*, 172–176.
15. Yao, J.; Yang, S.; Lin, X.; Peng, Y. Incremental double sequence algorithm of concept lattice union. *Appl. Res. Comput.* **2013**, *30*, 1038–1040.
16. Zhang, W.; Wei, L.; Qi, J. Attribute Reduction Theory and Method of Concept Lattice. *Sci. China Ser. E Inf. Sci.* **2005**, *35*, 628–639.
17. Wang, X.; Zhang, W. Attribute reduction in concept lattice and attribute characteristics. *Comput. Eng. Appl.* **2008**, *44*, 1–4.
18. Hu, L.; Zhang, J.; Zhang, S. Pruning based incremental construction of concept lattice. *J. Comput. Appl.* **2006**, *26*, 1659–1661.
19. Liu, B.; Zhang, C. New concept lattice structure-interval concept lattice. *Comput. Sci.* **2012**, *39*, 273–277.
20. Wang, L. *Research of Effective Construction Algorithm for Interval Concept Lattice and Its Application*; North China University of Science and Technology: Tangshan, China, 2015.
21. Zhang, C.; Wang, L.; Liu, B. Dynamic reduction theory for interval concept lattice based on covering and its realization. *J. Shandong Univ. (Nat. Sci.)* **2014**, *49*, 15–21.
22. Zhang, C.; Wang, L.; Liu, B. Longitudinal maintenance theory and algorithm for interval concept lattice. *Comput. Eng. Sci.* **2015**, *37*, 1221–1226.

23. Wang, Z.; Hu, K.; Hu, X.; Liu, Z.; Zhang, D. General and incremental algorithms of rule extraction based on concept lattice. *Chin. J. Comput.* **1999**, *22*, 167–171.

24. Sahami, M. Learning classification rules using lattices. In *European Conference on Machine Learning*; Springer: Berlin/Heidelberg, Germany, 1995; pp. 343–346.

25. Liang, J.; Wang, J. An algorithm for extracting rule-generating sets based on concept lattice. *J. Comput. Res. Dev.* **2004**, *41*, 1339–1344.

26. Xie, Z.; Liu, Z. A fast incremental algorithm for building concept lattice. *Chin. J. Comput.* **2002**, *25*, 490–496.

27. Hu, K.; Lu, Y.; Shi, C. An Integrated Mining Approach for Classification and Association Rule Based on Concept Lattice. *J. Softw.* **2000**, *11*, 1478–1484.

28. Wang, W. *Research on Association Rule Mining and Change Model Based on Concept Lattice*; Shandong University: Jinan, China, 2012.

29. Wang, L.; Zhang, C.; Liu, B. Mining Algorithm of Interval Association Rule with Parameters and Its Application. *J. Front. Comput. Sci. Technol.* **2016**, *10*, 1546–1554.

30. Xu, J.; Li, M.; Liu, B. Rule optimization method and application based on interval concept lattice. *Comput. Eng. Appl.* **2017**, *53*, 167–173.

31. Zhang, R.; Zhang, C.; Wang, L.; Liu, B. Vertical union algorithm of interval concept lattices. *J. Comput. Appl.* **2015**, *35*, 1217–1221.

32. Yang, Y.; Zhang, R.; Liu, B. Dynamic Horizontal Union Algorithm for Multiple Interval Concept Lattices. *Mathematics* **2019**, *7*, 159. [CrossRef]

33. Liu, B.; Zhang, C.; Yang, Y. Research on Interval Concept Lattice and its Construction Algorithm. *Prz. Elektrotechniczny* **2013**, *2013*, 1.

 *mathematics*

Article

On the Efficacy of Ensemble of Constraint Handling Techniques in Self-Adaptive Differential Evolution

Hassan Javed [1,†], Muhammad Asif Jan [1,*,†], Nasser Tairan [2,†], Wali Khan Mashwani [1,†], Rashida Adeeb Khanum [3,†], Muhammad Sulaiman [4,†], Hidayat Ullah Khan [5,†] and Habib Shah [2,†]

1 Institute of Numerical Sciences, Kohat University of Science & Technology, Kohat 26000, Pakistan
2 College of Computer Science, King Khalid University, Abha 61321, Saudi Arabia
3 Jinnah College for Women, University of Peshawar, Peshawar 25000, Pakistan
4 Department of Mathematics, Abdul Wali Khan University, Mardan 23200, Pakistan
5 Department of Economics, Abbottabad University of Science & Technology, Abbottabad 22010, Pakistan
* Correspondence: majan.math@gmail.com or majan@kust.edu.pk
† These authors contributed equally to this work.

Received: 9 June 2019; Accepted: 10 July 2019; Published: 17 July 2019

Abstract: Self-adaptive variants of evolutionary algorithms (EAs) tune their parameters on the go by learning from the search history. Adaptive differential evolution with optional external archive (JADE) and self-adaptive differential evolution (SaDE) are two well-known self-adaptive versions of differential evolution (DE). They are both unconstrained search and optimization algorithms. However, if some constraint handling techniques (CHTs) are incorporated in their frameworks, then they can be used to solve constrained optimization problems (COPs). In an early work, an ensemble of constraint handling techniques (ECHT) is probabilistically hybridized with the basic version of DE. The ECHT consists of four different CHTs: superiority of feasible solutions, self-adaptive penalty, ε-constraint handling technique and stochastic ranking. This paper employs ECHT in the selection schemes, where offspring competes with their parents for survival to the next generation, of JADE and SaDE. As a result, JADE-ECHT and SaDE-ECHT are developed, which are the constrained variants of JADE and SaDE. Both algorithms are tested on 24 COPs and the experimental results are collected and compared according to algorithms' evaluation criteria of CEC'06. Their comparison, in terms of feasibility rate (FR) and success rate (SR), shows that SaDE-ECHT surpasses JADE-ECHT in terms of FR, while JADE-ECHT outperforms SaDE-ECHT in terms of SR.

Keywords: evolutionary algorithms; formal methods in evolutionary algorithms, differential evolution, self-adaptive differential evolutionary algorithms; metaheuristics; mutation strategies; parameters' adaptation; constrained optimization; ensemble of constraint handling techniques; and hybrid algorithms

---

1. Introduction

Evolutionary algorithms (EAs) are nature inspired population-based stochastic search and optimization methods. EAs work on the principle of natural evolution. In EAs, selected population members based on a fitness/selection scheme, the so called parents, undergo perturbation by applying genetic operators, mutation and crossover, to produce offspring. A selection scheme is then adopted to select the fittest individuals with a certain probability among the parents and offspring for the next generation. Many EAs, such as genetic algorithms (GAs), differential evolution (DE), particle swarm optimization (PSO), firefly algorithm (FA), bee algorithm (BA), ant colony optimization (ACO), evolution strategy (ES) etc. have been designed using different genetic operators and selection schemes for the unconstrained optimization problems since 1959.

Differential evolution (DE) [1] has proven to be a simple and efficient EA for many optimization problems. A number of variants of DE were developed and are in practice for unconstrained/constrained optimization [2–9]. In DE, a random initial population of size NP is generated in the whole search space to a possible extent and the fittest/best with minimum function value in the initial population is found. It then invokes one of the different mutation strategies such as DE/rand/1, DE/current to best/2, DE/rand/2, DE/current-to-rand/1 to generate a mutant vector. For example in DE/rand/1, the weighted difference scaled by a scaling factor $F \in [0,\ 2]$ between two population vectors is added to a target vector to generate a mutant vector. Afterwards, the parameters of mutant vector and target vector are mixed with a certain crossover probability $C_r \in [0,\ 1]$ to produce the trial vector. Using the one-to-one-spawning selection mechanism, if the objective function value of the trial vector is less than the objective function value of the target vector, in minimization sense, then the trial vector replaces the target vector and becomes the parent for the next generation. The three steps of producing the mutant vector, the trail vector and comparison of the target and trial vectors are repeated until a stopping criterion is met. Also, the fittest/best individual is updated after every generation by comparing the function values of the trial vector, if it is successful in selection process, and fittest/best individual found so far. For more details of DE and different mutation strategies used in it, the readers are referred to [10,11].

The performance of the original DE algorithm is highly dependent on the mutation strategies and its parameters' settings [11–14]. During different evolution stages, different strategies and different parameters' settings with different global and local search capabilities might be preferred. Huang et al. developed a self-adaptive DE variant, SaDE [10]. SaDE automatically adapts the learning strategies and the parameters' settings during evolution. It probabilistically selects one of the four mutation strategies: DE/rand/1, DE/current to best/2, DE/rand/2, DE/current-to-rand/1 for each individual in the current population. J. Zhang and A. C. Sanderson developed another self-adaptive DE version, self-adaptive differential evolution with optional external archive (JADE) [15]. JADE too automates the parameters and employs the mutation strategy DE/current-to-pbest with the optional external archive. The strategy DE/current-to-pbest uses not only the information of the best solution, but also the information on the other good solutions. The external archive keeps record of the inferior solutions, which are then used for diversity among population members and avoiding premature convergence.

For recent advances in DE, the readers are referred to [16,17]. EAs suit a variety of applications in the fields of engineering and science [18–24]. Generally, EAs outperform traditional optimization algorithms for problems which are not continuous, non-differentiable, multi-modal, noisy and not well-defined. However, EAs are unconstrained optimization techniques. They are not capable to directly solve COPs having constraints of any kind (e.g., equality, inequality, linear and non-linear etc.). To overcome this problem, CHTs are used with EAs to handle all types of constraints. The last three decades have witnessed many techniques for handling constraints by EAs [20,25]. Michalewicz and Schoenauer [26] categorized them into five classes: preserving feasibility of solutions, adopting penalty functions, separating feasible solutions from infeasible ones, decoding, and hybridizing different techniques. However, according to no free lunch theorem (NFL) [27], a single CHT can not outperform all other CHTs on each problem. Same is true for different EAs as well. Thus, one has to try and combine different CHTs and EAs to design a suitable algorithm that can solve most of the problems. So keeping in mind the NFL theorem and some other individual problems of COPs, an ensemble of constraint handling techniques (ECHT) is combined with the basic version of DE in [28,29]. ECHT consists of four different CHTs: superiority of feasible solutions, self-adaptive penalty, ε-constraint handling technique and stochastic ranking. SaDE and JADE, being advanced self-adaptive variants, are both unconstrained search and optimization algorithms. Like other EAs, they also need some additional CHTs to solve constrained optimization problems (COPs).

In this work, the ECHT is implemented in the selection scheme, where offspring and parents compete for survival to next generation, of JADE and SaDE. As a result, constrained versions of JADE and SaDE, denoted by JADE-ECHT and SaDE-ECHT, are developed. The performance of JADE-ECHT and SaDE-ECHT is tested and compared based on feasibility rate (FR) and success rate (SR) on 24 COPs according to algorithms' evaluation criteria of CEC'06.

This rest of this paper is ordered as follows. The general COP and ECHT are detailed in Section 2. Section 3 presents the proposed modified algorithms, JADE-ECHT and SaDE-ECHT. Section 4 presents and discusses the experimental results obtained with JADE-ECHT and SaDE-ECHT. Finally, Section 5 describes the concluding remarks of this work.

2. Constrained Optimization Problem and ECHT

This section first describes the constrained optimization problem to be considered in this work. It then illustrates the four CHTs of ECHT.

2.1. Constrained Optimization Problem (COP)

Time, physical, and geometric etc. type constraints exist in most of the real world optimization problems. Such problems can be modelled as a COP. Mathematically, a COP, in case of minimization, can be formulated as follows [30]:

$$
\begin{aligned}
&\text{Minimize} f(\mathbf{x}), \ \mathbf{x} = [x_1, x_2, \ldots, x_n]^T \\
&\text{subject to} \\
&g_i(\mathbf{x}) \leq 0 \qquad i = 1, \ldots, l, \\
&h_j(\mathbf{x}) = 0 \qquad j = l+1, \ldots, p, \\
&l_i \leq x_i \leq u_i, i = 1, 2, \ldots, n..
\end{aligned}
\tag{1}
$$

In problem (1), $f(\mathbf{x})$ is called cost function which will be minimized. In case of maximizing the cost function, it needs to be multiplied with a negative sign. The n-dimensional vector $\mathbf{x}$ is called decision variable vector. There are l inequality and $p - l$ equality constraints. An inequality constraint $g_j(\mathbf{x})$ becomes an active constraint, if $g_j(\mathbf{x}^*) = 0$, where $\mathbf{x}^*$ is global optimum solution, whereas equality constraints, $h_j(\mathbf{x}) = 0$, are active by default. Generally, equality constraints are converted into inequality constraints by $|h_j(\mathbf{x})| - \epsilon \leq 0$, where ϵ is an acceptable tolerance for equality constraints. According to CEC'06 [30] evaluation criteria, ϵ is set to 0.0001 (in this work, we will also use the same value for ϵ). l_i and u_i are the lower and upper bounds of component x_i of vector $\mathbf{x}$. They form the whole search space S. The solution $\mathbf{x} \in S$ is referred to be feasible, if it satisfies all the equality and inequality constraints of problem (1); otherwise, it is called infeasible. We denote with F the set of all feasible solutions and normally $F \subset S$. The total constraints' violation for an infeasible solution is defined as [28,29]:

$$
v(\mathbf{x}) = \frac{\sum_{i=1}^{p} c_i(g_i'(\mathbf{x}))}{\sum_{i=1}^{p} c_i},
\tag{2}
$$

where

$$
g_i'(\mathbf{x}) = \begin{cases} max\{g_i(\mathbf{x}), 0\}, & i = 1, \ldots, l \\ max\{|h_j(\mathbf{x})| - \epsilon, 0\}, & j = l+1, \ldots, p. \end{cases}
\tag{3}
$$

where $c_i (= 1/g_{max_i}')$ denotes weight parameter, g_{max_i}' denotes the maximum constraint violation of constraint $g_i(\mathbf{x}), i = 1, \ldots, l$ obtained thus far. It maybe noted that c_i changes during the evolution process. This helps in balancing how each constraint contributes in the problem irrespective of their different numerical ranges. The four constraints handling techniques which are used in this work are detailed as follows.

2.2. Superiority of Feasible Solutions (SF)

As the name suggests, in SF feasible solutions have priority over infeasible solutions. SF was first suggested by Deb [31]. In this method, two solutions, a parent $\mathbf{x}^i$ and an offspring $\mathbf{x}^j$ compete. The parent $\mathbf{x}^i$ is considered better than the offspring $\mathbf{x}^j$, if any of the subsequent three settings is met [31]:

- Parent, $\mathbf{x}^i$ is feasible and offspring, $\mathbf{x}^j$ is infeasible.
- Both parent and offspring, $\mathbf{x}^i$ and $\mathbf{x}^j$ are feasible, but parent, $\mathbf{x}^i$ has minimum fitness value than the offspring, $\mathbf{x}^j$.
- Both $\mathbf{x}^i$ and $\mathbf{x}^j$ are infeasible, and overall constraints' violation $v(\mathbf{x}^i)$ of parent, $\mathbf{x}^i$ is less than overall constraints' violation $v(\mathbf{x}^j)$ of offspring, $\mathbf{x}^j$, where $v(\mathbf{x}^i)$ and $v(\mathbf{x}^j)$ are calculated by using Equation (2).

2.3. Self-Adaptive Penalty (SP)

Penalty methods are the most common approaches to handle constraints in the family of CHTs. In these techniques, in order to penalize an infeasible solution, the cost value of each infeasible solution and a penalty term corresponding to its constraints' violation are added, in minimization sense (subtracted in maximization sense). In SP [28], an attempt has been made to facilitate the algorithm to search for feasible solutions, in case there are few feasible solutions, and find the optimum, in case there are enough feasible solutions. For this purpose, two penalties are added to the cost of an infeasible solution. This help in identifying the best infeasible solutions in the existing population. The amount of the added penalties considers the number of feasible solutions that exist in the current population. Thus, if there are few feasible solutions in the combined population of parents and offspring, the amount of penalty to infeasible individuals with higher constraints' violation will be greater. On the contrary, with many feasible solutions, the fittest infeasible solutions in terms of cost are less penalized.

2.4. The ε-Constraint (EC) Handling Technique

The ε-constraint (EC) handling technique [32] adopts the parameter ε to relax the active constraints. The parameter ε is updated until a fixed generation counter is reached. Afterwards, ε becomes 0 to get individuals with no constraints' violation (for detailed formulation of this technique, please see [28,32]).

2.5. Stochastic Ranking (SR)

SR [33] stochastically balances overall constraints' violation and fitness function value. A solution is ranked based on its cost value, if it is feasible or if a randomly generated number is smaller than a probability factor p_f; otherwise, it is ranked on the constraints' violation. The proposed value of $p_f = 0.475$. However, if this constant value is not used, then it decreases linearly from $p_f = 0.475$ to $p_f = 0.025$ from initial generation to the last generation.

In [28], the ECHT is tested with evolutionary programming (EP) and basic DE. In this paper, we hybridize ECHT with the advanced versions of DE, JADE [15] and SaDE [10]

3. JADE-ECHT and SaDE-ECHT

In this section, we first give the algorithmic details of JADE-ECHT, which is then followed by the details of SaDE-ECHT.

3.1. JADE-ECHT

JADE [15] is an updated version of DE. It is also an unconstrained optimization algorithm. So it needs some additional CHTs to solve COPs. In this work, we embed the four above discussed CHTs in the selection scheme of JADE to modify it for solving COPs. The whole procedure of the proposed technique JADE-ECHT, shown in Figure 1, is discussed as follows.

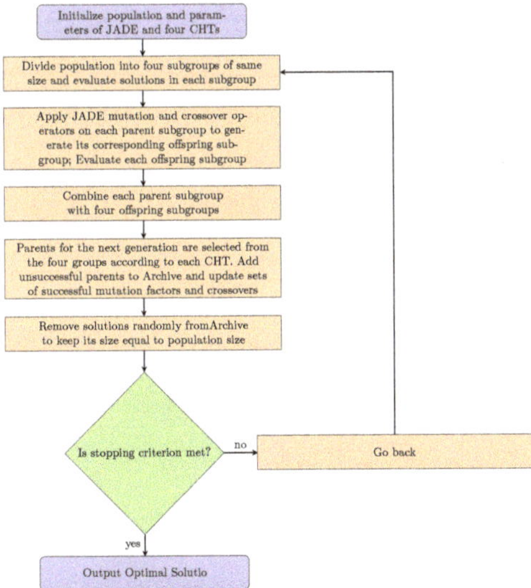

Figure 1. Flowchart of self-adaptive differential evolution with optional external archive (JADE)-ensemble of constraint handling techniques (ECHT).

Step 1: generate initial population P, set the generation number $t = 1$, initial crossover probability $\mu_{CR} = 0.5$, initial mutation factor $\mu_F = 0.5$, the set of archive inferior solutions $A_i = \varnothing$, the sets of successful mutation factors and crossovers, $S_F^i = \varnothing$, $S_{CR}^i = \varnothing$, respectively, where $i = 1, \ldots, 4$.

Step 2: divide population P into four subpopulations, $P_i, i = 1, \ldots, 4$ each of size PS (population size to be tackled by each CHT). Set parameters $PAR_i, i = 1, \ldots, 4$ of PS individuals each with dimension D according to the rules of JADE and corresponding CHT. Also, calculate F_l^i and $CR_l^i, \forall l \in \{1, \ldots, PS\}$, where $CR_l^i = randn_l^i(\mu_{CR}, 0.1)$, $F_l^i = randn_l^i(\mu_{CR}, 0.1)$.

Step 3: compute the cost and the total constraints' violation for every solution in each subpopulation using Equations (1)–(3).

Step 4: each parent subpopulation $(P_i, i = 1, \ldots, 4)$ generates offspring subpopulation $(OFF_i, i = 1, \ldots, 4)$ as a result of applying mutation and crossover operators, respectively as follows [15]:

$$\mathbf{v}_{l,t}^i = \mathbf{x}_{l,t}^i + F_l^i \cdot (\mathbf{x}_{pbest,t}^i - \mathbf{x}_{l,t}^i) + F_l^i \cdot (\mathbf{x}_{r_1^i,t}^i - \tilde{\mathbf{x}}_{r_2^i,t}^i),$$

where $\mathbf{x}_{pbest,t}$ is one of the 100P% best vectors. and $\tilde{\mathbf{x}}_{r_2^i,t}^i \neq \mathbf{x}_{r_1^i,t}^i \neq \mathbf{x}_{l,t}^i$ are chosen randomly from the existing population, P and from the union of current population and archived population, $P \cup A$. The archive A retains the parent individuals that are unsuccessful in the selection scheme.

$$u_{l,t}^i = \begin{cases} v_{l,t}^i, & \text{if } (rand_j[0,1] < \mu_{CR} \text{ or } (j = j_{rand}) \\ x_{l,t}^i, & \text{otherwise} . \end{cases} \tag{4}$$

In Equation (4), $v_{l,t}^i$ and $x_{l,t}^i$ are the lth components of the ith mutant and trial vectors in generation t.

Step 5: evaluate the cost and the total constraints' violation for every offspring in each subpopulation using Equations (1)–(3). Every offspring holds the cost and constraints values distinctly.

Step 6: each parent subpopulation is grouped together with its own offspring and the offspring produced by the remaining three subpopulations corresponding to different CHTs. This way four different groups of populations are generated.

Step 7: Parents population P for the next generation is selected from the four groups according to the rule of each CHT. Unsuccessful parents are added to the archive A_i. All successful crossover probabilities from CR^i_{PS} and mutation factors from F^i_{PS} are added to S^i_{CR} and S^i_F.

Step 8: Remove solutions randomly from A_i so that $|A_i| \leq PS$. Update μ_{CR} and μ_F adopting the formulations of [28].

Step 9: If the stopping criteria are not met, go to Step 2; otherwise, stop.

3.2. SaDE-ECHT

SaDE [10] is also an unconstrained optimization algorithm. Like JADE and other EAs, it also needs some additional mechanisms to solve COPs. In this work, the four CHTs of ECHT are used in the selection scheme of SaDE for solving COPs. The whole procedure of proposed SaDE-ECHT, shown in Figure 2, is as follow:

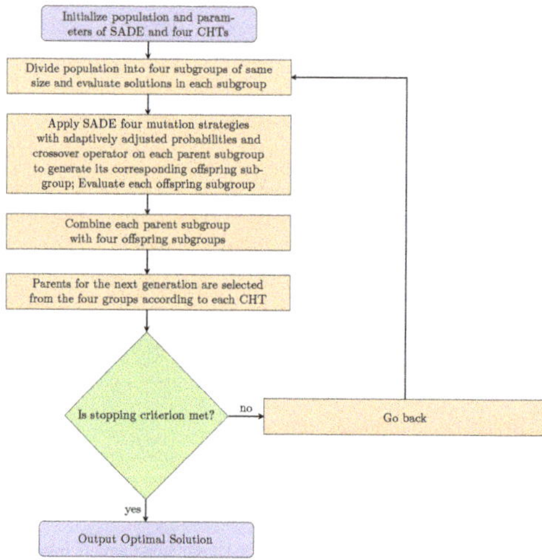

Figure 2. Flowchart of JADE-ECHT.

Step 1: generate initial population P, initiate the generation counter $t = 1$, initial crossover probability $\mu_{CR} = 0.5$, initial mutation factor $\mu_F = 0.5$.

Step 2: divide population P into four subpopulations, $P_i, i = 1, \dots, 4$ each of size PS. Set parameters PAR_i, where $i = 1, \dots, 4$ of PS individuals each with dimension D and generate F in [0,2] and CR in (0,1) by using normal distribution according to the rules of SaDE and corresponding CHT.

Step 3: compute the cost and the total constraints' violation for every solution in each subpopulation using Equations (1)–(3).

Step 4: each parent in each subpopulation produces offspring by using one of the four mutation strategies, DE/rand/1, DE/current-to-best/2, DE/rand/2, and DE/current-to-rand/1 (for details of these strategies, please see [10]) and crossover given in Equation (4). For first 20 generations, probabilities are fixed and set to $p_1 = p_2 = p_3 = p_4 = 0.25$. Afterwards, the Roulette Wheel selection is adopted to update the respective probability p_i as follows [10]:

$$p_i = \frac{ns_i}{ns_i + nf_i}, i = 1, 2, 3, 4 \tag{5}$$

Step 5: evaluate the cost and the total constraints' violation for every offspring in each subpopulation using Equations (1)–(3).

Step 6: each parent subpopulation is grouped together with its own offspring and the offspring produced by the remaining three subpopulations corresponding to different CHTs. This way four different groups of populations are generated.

Step 7: parents population P for the next generation are selected from the four groups according to the rule of each CHT.

Step 8: recalculate crossover probability after every five generations according to the mean of recorded CR values.

Step 9: if the stopping criteria are not met, go to Step 2; otherwise, stop.

4. Experimental Results

The performances of JADE-ECHT and SaDE-ECHT were evaluated on the suit of CEC'06, which contains twenty four benchmark functions. The PC configuration and parameters' settings are given in Tables 1 and 2.

Table 1. Configuration of the PC.

System	Windows 8
CPU	3.00 GHz
Ram	2 GB
Language	MATLAB 2012, 8.0.0.783

Table 2. Parameters' settings.

Parameters' Description	Parameters' Settings
Population size for each CHT	$PS = 25$
Whole population size	$NP = 4 * PS = 100$
Maximum number of generations	$t = 2500$
Total number of runs	$runs = 25$
Initial value of mutation factor	$\mu_F = 0.5$
Initial value of crossover probability	$\mu_{CR} = 0.5$
Termination criterion based on maximum function evaluations	$max\_FEs = 500{,}000.$

4.1. Result Achieved

In Tables 3–6, a comparison of both algorithms after 5×10^5 FEs is shown. All the obtained results are gathered according to CEC'06 [30] algorithms' evaluation criteria for problems g01 to g24. The criteria include collecting statistics of the best (minimum), worst (maximum), median, mean and standard deviation of the function error values $f(\mathbf{x}) - f(\mathbf{x}^*)$, where $f(\mathbf{x})$ is the best objective function value obtained by the algorithm after 5×10^5 FEs and $f(\mathbf{x}^*)$ is the know objective function value at the optimal solution. The numbers in parenthesis after the objective function value show the number of violated constraints, whereas c determines the number of violated constraints at the median solution with violation greater than $0.1, 0.001, 0.0001$. $\bar{v}$ shows mean violation at median solution, FR is the feasibility rate which is defined as the number of feasible runs over total runs, and SR is success rate given by the number of successful runs over total runs. A run is called a feasible run, if the algorithm attains in max_FEs at least one feasible solution. Likewise, a run is successful, if the algorithm gets a feasible solution for which the function error value is smaller than 0.0001 in max_FEs.

Table 3 compares the experimental results achieved by JADE-ECHT and SaDE-ECHT for problems g01–g06. This table shows that SaDE-ECHT achieved better statistics in terms of best, median, mean and standard deviation values than JADE-ECHT on problems g01 and g03, whereas JADE-ECHT surpasses SaDE-ECHT on problems g02 and g05 except the best value of g02. It can also be observed from the same table that both algorithms show comparable performance on problems g04 and g06.

The table also shows that both algorithms have achieved 100% FR on all six problems, as can be confirmed from the 0s in parenthesis after the objective function values, and columns for c and $\bar{v}$. The SR of SaDE-ECHT on problems g01–g03 is higher than JADE-ECHT. JADE-ECHT's SR is better than SaDE-ECHT on problem g05, while both algorithms obtained the same SR of 100% on problems g04 and g06.

Table 3. Comparison of self-adaptive differential evolution with optional external archive (JADE)-ensemble of constraint handling techniques (ECHT) and self-adaptive differential evolution (SaDE)-ECHT after FES = 500,000 for g01–g06. The bold numbers indicate the better results.

Prob	Algorithm	Best	Median	Worst	c	$\bar{v}$	Mean	Std	FR	SR
g01	JADE-ECHT	0(0)	0(0)	2.0000(0)	0,0,0	0	0.0800	0.4000	100%	96%
	SaDE-ECHT	0(0)	0(0)	**0(0)**	0,0,0	0	**0**	**0**	100%	**100%**
g02	JADE-ECHT	0.0001(0)	**0.0004(0)**	**0.0276(0)**	0,0,0	0	**0.0064**	**0.0091**	100%	16%
	SaDE-ECHT	**0(0)**	0.0110(0)	0.1263(0)	0,0,0	0	0.0191	0.0254	100%	**24%**
g03	JADE-ECHT	0.0250(0)	0.1015(0)	0.4245(0)	0,0,0	0	0.1385	0.1036	100%	0%
	SaDE-ECHT	**0(0)**	**0.0122(0)**	**0.1524(0)**	0,0,0	0	**0.0243**	**0.0343**	100%	**12%**
g04	JADE-ECHT	0(0)	0(0)	0(0)	0,0,0	0	0	0	100%	100%
	SaDE-ECHT	0(0)	0(0)	0(0)	0,0,0	0	0	0	100%	100%
g05	JADE-ECHT	0(0)	**0(0)**	**0(0)**	0,0,0	0	**0**	**0**	100%	**100%**
	SaDE-ECHT	0(0)	91.4773(0)	515.4900(0)	0,0,0	0	110.1546	101.2496	100%	4%
g06	JADE-ECHT	0(0)	0(0)	0(0)	0,0,0	0	0	0	100%	100%
	SaDE-ECHT	0(0)	0(0)	0(0)	0,0,0	0	0	0	100%	100%

Table 4 presents the experimental statistics achieved by JADE-ECHT and SaDE-ECHT for problems g07–g12. The results of this table show that both algorithms obtained comparable statistics for problems g08, g11 and g12. This table also shows superior performance of SaDE-ECHT in terms median, mean and standard deviation values than JADE-ECHT on the problems g07, g09 and g10 except the best values on problems g07 and g10, where JADE-ECHT got better best values. The table also confirms that both algorithms have achieved 100% FR on all six problems, as can be seen from the 0s in parenthesis after the objective function values, and columns for c and $\bar{v}$. The SR of JADE-ECHT on problems g07 and g10 is higher than SaDE-ECHT. SaDE-ECHT's SR is better than JADE-ECHT on problem g09, while both algorithms obtained the same SR of 100% on problems g08, g11 and g12.

Table 5 demonstrates the experimental results achieved by JADE-ECHT and SaDE-ECHT for problems g13–g18. The results of this table show that both algorithms performed similar on problem g16. This table also shows superior performance of SaDE-ECHT in terms best, median, mean and standard deviation values than JADE-ECHT on problems g13 and g18 except the standard deviation of g13, while JADE-ECHT performed better than SaDE-ECHT on problems g14, g15 and g17 except the mean and standard deviation values of problem g14, where SaDE-ECHT got better values for the two quantities. The table also confirms that both algorithms have achieved 100% FR on all six problems, as can be seen from the 0s in parenthesis after the objective function values, and columns for c and $\bar{v}$. The SR of JADE-ECHT on problems g14, g15, and g17 is higher than SaDE-ECHT. SaDE-ECHT's SR is better than JADE-ECHT on problems g13 and g18, while both algorithms obtained the same SR of 100% on problem g16.

Table 4. Comparison of JADE-ECHT and SaDE-ECHT after FES = 500,000 for g07–g12. The bold numbers indicate the better results.

Prob	Algorithm	Best	Median	Worst	c	$\bar{v}$	Mean	Std	FR	SR
g07	JADE-ECHT	**0**(**0**)	0.0879(0)	**0.2651**(**0**)	0,0,0	0	0.0976	**0.0726**	100%	**4%**
	SaDE-ECHT	0.0001(0)	**0.0114**(**0**)	0.3230(0)	0,0,0	0	**0.0518**	0.0850	100%	0%
g08	JADE-ECHT	0(0)	0(0)	0(0)	0,0,0	0	0	0	100%	100%
	SaDE-ECHT	0(0)	0(0)	0(0)	0,0,0	0	0	0	100%	100%
g09	JADE-ECHT	0(0)	0.0039(0)	0.0714(0)	0,0,0	0	0.0132	0.0192	100%	20%
	SaDE-ECHT	0(0)	**0**(**0**)	**0.0006**(**0**)	0,0,0	0	**0.0001**	**0.0002**	100%	**76%**
g10	JADE-ECHT	**0**(**0**)	133.9677(0)	343.5425(0)	0,0,0	0	143.0809	105.4501	100%	**4%**
	SaDE-ECHT	0.0012(0)	**0.1709**(**0**)	**11.9004**(**0**)	0,0,0	0	**1.1748**	**3.0352**	100%	0%
g11	JADE-ECHT	0(0)	0(0)	0(0)	0,0,0	0	0	0	100%	100%
	SaDE-ECHT	0(0)	0(0)	0(0)	0,0,0	0	0	0	100%	100%
g12	JADE-ECHT	0(0)	0(0)	0(0)	0,0,0	0	0	0	100%	100%
	SaDE-ECHT	0(0)	0(0)	0(0)	0,0,0	0	0	0	100%	100%

Table 5. Comparison of JADE-ECHT and SaDE-ECHT after FES = 500,000 for g13–g18. The bold numbers indicate the better results.

Prob	Algorithm	Best	Median	Worst	c	$\bar{v}$	Mean	Std	FR	SR
g13	JADE-ECHT	0.3849(0)	0.9118(0)	0.9459(0)	0,0,0	0	0.8275	**0.1750**	100%	0%
	SaDE-ECHT	**0**(**0**)	**0.3870**(**0**)	**0.8491**(**0**)	0,0,0	0	**0.3608**	0.2828	100%	**4%**
g14	JADE-ECHT	**0**(**0**)	**0.0174**(**0**)	5.5402(0)	0,0,0	0	1.9415	2.2940	100%	**40%**
	SaDE-ECHT	0.4527(0)	1.6397(0)	**3.3912**(**0**)	0,0,0	0	**1.7600**	**0.6956**	100%	0%
g15	JADE-ECHT	0(0)	**0**(**0**)	**0**(**0**)	0,0,0	0	**0**	**0**	100%	**100%**
	SaDE-ECHT	0(0)	0.0009(0)	2.5449(0)	0,0,0	0	0.3333	0.6971	100%	44%
g16	JADE-ECHT	0(0)	0(0)	0(0)	0,0,0	0	0	0	100%	100%
	SaDE-ECHT	0(0)	0(0)	0(0)	0,0,0	0	0	0	100%	100%
g17	JADE-ECHT	**0**(**0**)	**0**(**0**)	**74.0580**(**0**)	0,0,0	0	**8.8870**	**24.5623**	100%	**88%**
	SaDE-ECHT	7.9251(0)	91.2351(0)	297.1687(0)	0,0,0	0	92.9967	50.4589	100%	0%
g18	JADE-ECHT	0(0)	0.0001(0)	0.0206(0)	0,0,0	0	0.0011	0.0041	100%	52%
	SaDE-ECHT	0(0)	**0**(**0**)	**0**(**0**)	0,0,0	0	**0**	**0**	100%	**100%**

Table 6 presents the experimental results achieved by JADE-ECHT and SaDE-ECHT for problems g19–g24. The results of this table show that both algorithms performed similar on problem g24. This table also shows superior performance of JADE-ECHT in terms best, median, mean and standard deviation values than SaDE-ECHT on problems g19, g20 g21 and g23, except the best value of problem g20 and standard deviation value of problem g23, while SaDE-ECHT performed better than JADE-ECHT on problem g22. The table also confirms that both algorithms have achieved 100% FR on problems g19 and g24, as can be seen from the $0s$ in parenthesis after the objective function values, and columns for c and $\bar{v}$. Both algorithms are unsuccessful in solving problems g20 and g20. The FR of JADE-ECHT on problem g21 is lower than SaDE-ECHT, while the situation is vice versa in case of SR. The FR and SR of JADE-ECHT on problem g23 is higher than SaDE-ECHT.

Table 6. Comparison of JADE-ECHT and SaDE-ECHT after FES = 500,000 for g19–g24. The bold numbers indicate the better results.

Prob	Algorithm	Best	Median	Worst	c	$\bar{v}$	Mean	Std	FR	SR
g19	JADE-ECHT	**0(0)**	**1.4028(0)**	**3.6498(0)**	0, 0, 0	0	**1.5502**	**1.0136**	100%	12%
	SaDE-ECHT	0.3671(0)	1.7022(0)	6.6604(0)	0, 0, 0	0	2.3120	1.9699	100%	0%
g20	JADE-ECHT	3.2029(9)	**6.2057(8)**	**15.4062(12)**	1, 1, 2	**1.1209**	**7.2582**	**3.5087**	0%	0%
	SaDE-ECHT	**2.4461(11)**	14.8045(9)	18.3511(11)	2, 4, 4	3.1946	13.1617	4.8304	0%	0%
g21	JADE-ECHT	0(0)	**0.0633(0)**	263.7866(1)	0, 0, 0	0	**39.1073**	63.9006	96%	**44%**
	SaDE-ECHT	0(0)	77.3185(0)	**110.2441(0)**	0, 0, 0	0	71.8631	**25.3368**	100%	4%
g22	JADE-ECHT	390.4334(4)	10,565.5111(3)	19,715.2233(4)	3, 3, 3	17,5401.6096	10,557.6213	6162.3243	0%	0%
	SaDE-ECHT	**292.6511(3)**	**8834.7836(3)**	**19,258.8965(3)**	3, 3, 3	**90,196.1317**	**9289.3437**	**4998.2886**	0%	0%
g23	JADE-ECHT	**0(0)**	8.5726(0)	601.1293(0)	0, 0, 0	0	**117.5730**	198.2664	**36%**	**36%**
	SaDE-ECHT	182.7482(0)	357.7081(0)	**518.9083(0)**	0, 0, 0	0	344.3397	**87.4764**	0%	0%
g24	JADE-ECHT	0(0)	0(0)	0(0)	0, 0, 0	0	0	0	100%	100%
	SaDE-ECHT	0(0)	0(0)	0(0)	0, 0, 0	0	0	0	100%	100%

Figure 3 compares the convergence graphs of JADE-ECHT and SaDE-ECHT for problems g01–g06. This figure shows that JADE-ECHT converges faster than SaDE-ECHT on problems g01, g05 and g06, as less number of *FEs* have been used by it. In case of problem g04, the convergence of SaDE-ECHT is speedy than JADE-ECHT, while in case of problems g02, g03 both algorithms converge at the same rate.

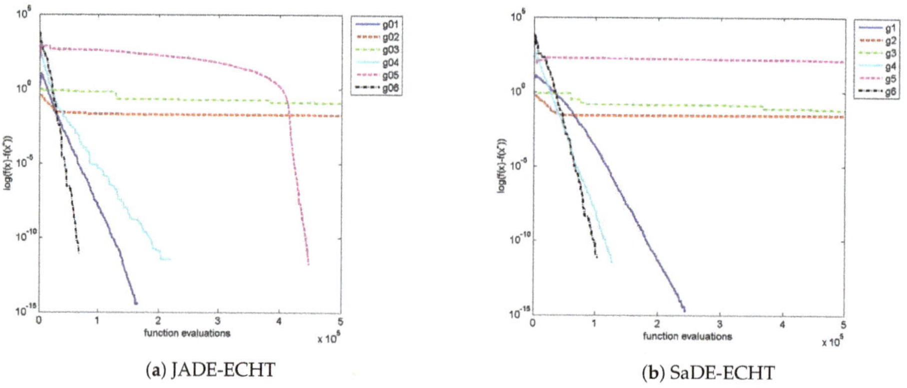

(a) JADE-ECHT   (b) SaDE-ECHT

Figure 3. Convergence comparison of JADE-ECHT and SaDE-ECHT for g01–g06.

Figure 4 compares the constraints' violations vs *FES* graphs of JADE-ECHT and SaDE-ECHT for problems g01–g06. This figure shows that both algorithms converge quickly to the feasible region and the optimal solution (s) thus has zero constraints' violations.

Figure 5 compares the convergence graphs of JADE-ECHT and SaDE-ECHT for problems g07–g12. This figure shows that both JADE-ECHT and SaDE-ECHT converge at the same rate for all six problems except g11, where JADE-ECHT converges faster than SaDE-ECHT.

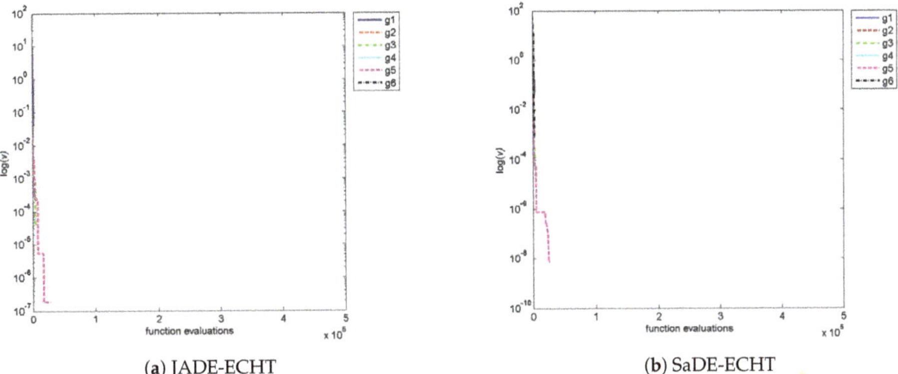

(a) JADE-ECHT   (b) SaDE-ECHT

Figure 4. Constraint violation comparison of JADE-ECHT and self-adaptive differential evolution (SaDE)-ECHT for g01–g06.

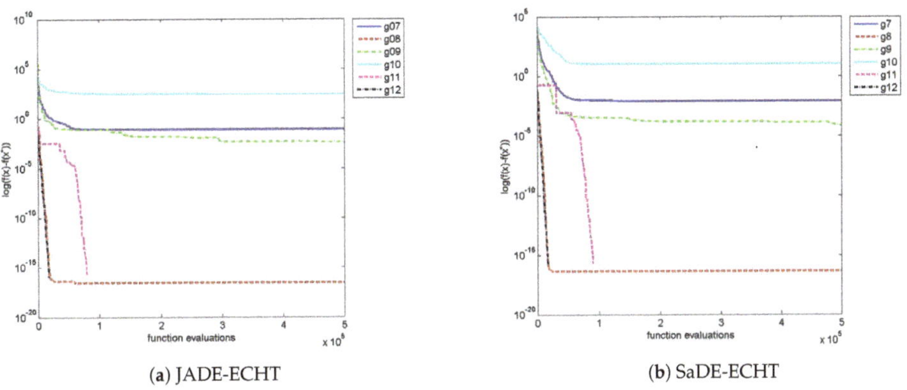

(a) JADE-ECHT   (b) SaDE-ECHT

Figure 5. Convergence comparison of JADE-ECHT and SaDE-ECHT for g07–g12.

Figure 6 compares the constraints' violations vs *FES* graphs of JADE-ECHT and SaDE-ECHT for problems g07–g12. This figure too shows that both algorithms converge quickly to the feasible region and optimal solution(s) thus has zero constraints' violations.

Figure 7 compares the convergence graphs of JADE-ECHT and SaDE-ECHT for problems g13–g18. This figure shows that both JADE-ECHT and SaDE-ECHT converge at the same rate for all six problems except g15, where JADE-ECHT converges faster than SaDE-ECHT.

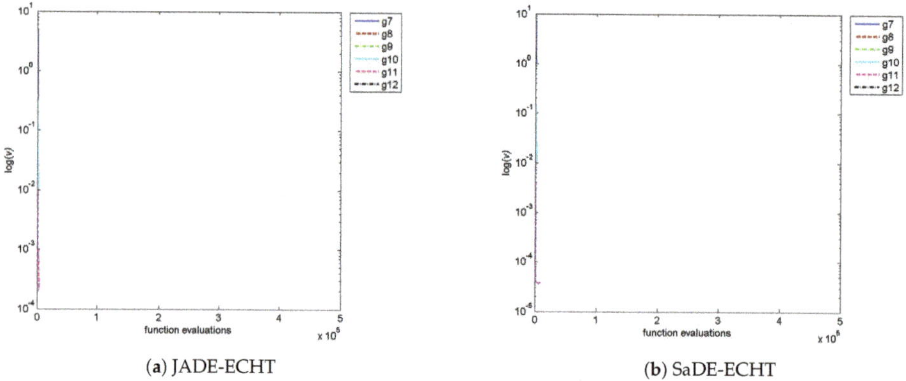

(**a**) JADE-ECHT

(**b**) SaDE-ECHT

Figure 6. Constraint violation comparison of JADE-ECHT and SaDE-ECHT for g07–g12.

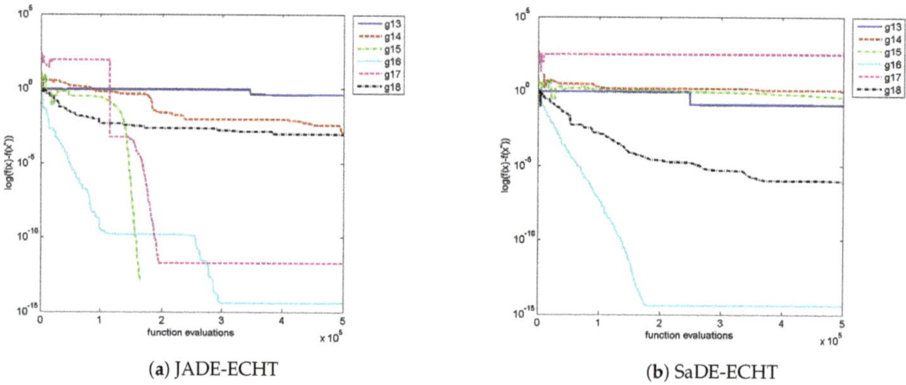

(**a**) JADE-ECHT

(**b**) SaDE-ECHT

Figure 7. Convergence comparison of JADE-ECHT and SaDE-ECHT for g13–g18.

Figure 8 compares the constraints' violations vs *FES* graphs of JADE-ECHT and SaDE-ECHT for problems g13–g18. This figure shows that both algorithms explore the infeasible region for about 1000 iterations and then converge to the feasible region. As a result, optimal solution(s) thus obtained has zero constraints' violations.

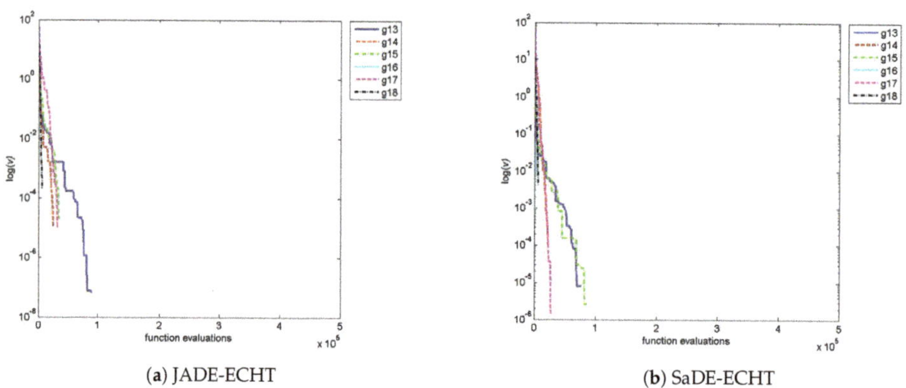

(**a**) JADE-ECHT

(**b**) SaDE-ECHT

Figure 8. Constraint violation comparison of JADE-ECHT and SaDE-ECHT for g13–g18.

Figure 9 compares the convergence graphs of JADE-ECHT and SaDE-ECHT for problems g19–g24. This figure shows that both JADE-ECHT and SaDE-ECHT converge almost at the same rate for all six problems and utilize the maximum function evaluations.

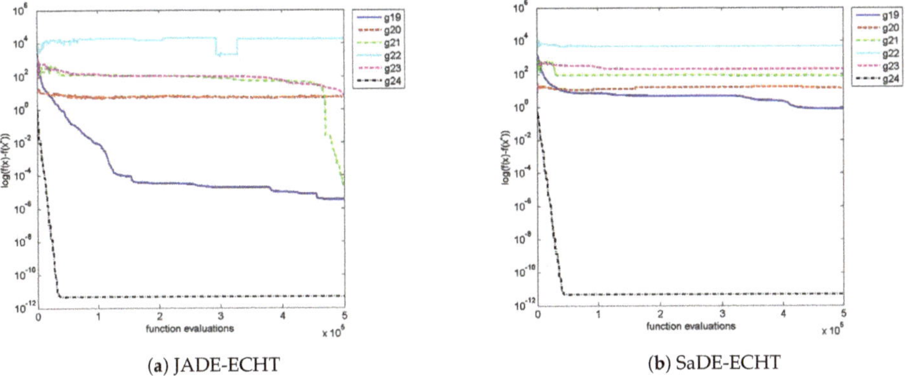

(a) JADE-ECHT   (b) SaDE-ECHT

Figure 9. Convergence comparison of JADE-ECHT and SaDE-ECHT for g19–g24.

Figure 10 compares the constraints' violations vs *FES* graphs of JADE-ECHT and SaDE-ECHT for problems g19–g24. This figure clearly shows that both algorithms failed to obtain any feasible solution in case of problems g20 and g22, although maximum function evaluations have been used.

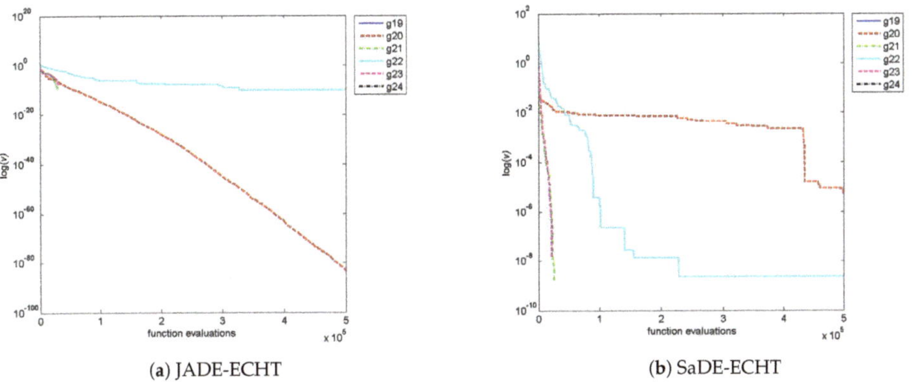

(a) JADE-ECHT   (b) SaDE-ECHT

Figure 10. Constraint violation comparison of JADE-ECHT and SaDE-ECHT for g19–g24.

Figures 3,5,7 and 9 show the comparison of the convergence graphs vs *FES* of both algorithms for all problems g01-g24, whereas Figures 4,6,8 and 10 demonstrate their comparison graphs of the constraints' violations vs *FEs*.

Overall, it can be concluded from the tabulated results and figures that both algorithms have achieved feasible solution (s) and near optimal solution (s) on 22 problems out of 24 except problems g20 and g22. The tables show that the FR of JADE-ECHT on 20 problems out of 24 is 100% and that of SaDE-ECHT on 22 problems out of 24 is 100%. The SR of JADE-ECHT on most of the problems is better than SaDE-ECHT. On two problems g20 and g22, the FR and SR of both algorithms are 0%. The dimension of these two problems is higher than other 22 problems. Also, these two problems had a large number of equality constraints. It can be noted from our experiments and some other literature review that equality constraints were hard to handle.

Table 7 compares the FR and SR of JADE-ECHT and SaDE-ECHT with other competing algorithms of CEC'2006. It can be seen from the said table that both JADE-ECHT and SaDE-ECHT achieved better FR, and can be placed at positions second and fourth, respectively. However, they failed to achieve better SR than the competing algorithms. A reason of failure could be the use of four different CHTs, where the resources (FEs) are distributed based on the success of each individual CHT, while the competing algorithms used just one CHT. The same can also be observed from Tables 8 and 9, where the median and standard deviation values obtained after 5×10^5 FEs of JADE-ECHT and SaDE-ECHT are compared with other competing algorithms (the values of the two quantities for the competing algorithms are taken from each source paper). Another reason of low SR could be observed from the figures showing constraints' violations vs FES graphs. It can be noticed from these graphs that both algorithms converge quickly to the feasible region. As a result, they less explore the infeasible region and consequently suffer from stagnation and premature convergence.

Table 7. Comparison of JADE-ECHT and SaDE-ECHT in terms of feasibility rate (FR) and success rate (SR) with algorithms of CEC 2006.

Algorithms	FR	SR
DE	95.65%	78.09%
DMS-PSO	100%	90.61%
ϵ DE	100%	95.65%
GDE	92.00%	77.39%
jDE-2	95.65%	80.00%
MDE	95.65%	87.65%
MPDE	94.96%	87.65%
PCX	95.65%	94.09%
PESO+	95.48%	67.83%
SaDE	100%	87.13%
JADE-ECHT	95.30%	57.04%
SaDE-ECHT	95.65%	46.43%

Table 8. Comparison of median values of JADE-ECHT, SaDE-ECHT and CEC'2006 algorithms achieved after 500,000 FEs. The bold numbers indicate the better results.

Prob	DE	DMS-PSO	ϵ DE	GDE	jDE-2	MDE	MPDE	PCX	PESO+	SaDE	JADE-ECHT	SaDE-ECHT
g01	**0(0)**	**0(0)**	**0(0)**	**0(0)**	**0(0)**	**0(0)**	**0(0)**	**0(0)**	**0(0)**	**0(0)**	**0(0)**	**0(0)**
g02	5.1700×10^{-8} (0)	**0(0)**	3.0933×10^{-8}(0)	2.3251×10^{-7}(0)	3.3051×10^{-9}(0)	0.017460(0)	3.5608×10^{-6}	**0(0)**	1.4314×10^{-6}(0)	3.0800×10^{-9}(0)	0.0004(0)	0.0110(0)
g03	6.7110×10^{-1}(0)	**0(0)**	-4.4409×10^{-16}(0)	9.3634×10^{-1}(0)	0.3481(0)	**0(0)**	-2.8866×10^{-15}	**0(0)**	1.5890×10^{-7} (0)	1.7770×10^{-8}(0)	0.1015(0)	0.0122(0)
g04	7.6398×10^{-11}(0)	**0(0)**	**0(0)**	8.0036×10^{-11}(0)	**0(0)**	**0(0)**	3.6380×10^{-12}	**0(0)**	1.0000×10^{-10}(0)	2.1667×10^{-7}(0)	**0(0)**	**0(0)**
g05	-9.0949×10^{-13}(0)	**0(0)**	**0(0)**	**0(0)**	**0(0)**	**0(0)**	**0(0)**	**0(0)**	**0(0)**	**0(0)**	**0(0)**	91.4773(0)
g06	4.5475×10^{-11}	**0(0)**	1.1823×10^{-11}(0)	6.1846×10^{-11}(0)	1.1823×10^{-11}(0)	**0(0)**	1.0914×10^{-11}	**0(0)**	1.0000×10^{-10}(0)	4.5475×10^{-11}(0)	**0(0)**	**0(0)**
g07	7.9783×10^{-11}(0)	**0(0)**	-1.8474×10^{-13}(0)	3.6402×10^{-10}(0)	-1.8829×10^{-13}(0)	**0(0)**	-1.8474×10^{-13}	**0(0)**	9.4367×10^{-6}(0)	1.4608×10^{-7}(0)	0.0879(0)	0.0114(0)
g08	8.1964×10^{-11}(0)	**0(0)**	4.1633×10^{-17}(0)	8.1964×10^{-11}(0)	4.1633×10^{-17}(0)	**0(0)**	4.1633×10^{-17}	**0(0)**	1.0000×10^{-10}(0)	8.1964×10^{-11}(0)	**0(0)**	**0(0)**
g09	-9.8112×10^{-11}(0)	**0(0)**	**0(0)**	-9.7884×10^{-11}(0)	2.2737×10^{-13}(0)	**0(0)**	1.1369×10^{-13}	**0(0)**	1.0000×10^{-10}(0)	3.7440×10^{-7}(0)	0.0039(0)	**0(0)**
g10	6.2755×10^{-11}(0)	1.0124×10^{-8}(0)	-9.0949×10^{-13}(0)	6.9122×10^{-11}(0)	-9.0949×10^{-13}(0)	**0(0)**	-9.0949×10^{-13}	**0(0)**	1.3432×10^{-3}(0)	1.8120×10^{-6}(0)	133.9677(0)	0.1709(0)
g11	**0(0)**	**0(0)**	**0(0)**	**0(0)**	**0(0)**	**0(0)**	**0(0)**	**0(0)**	**0(0)**	**0(0)**	**0(0)**	**0(0)**
g12	**0(0)**	**0(0)**	**0(0)**	**0(0)**	**0(0)**	**0(0)**	**0(0)**	**0(0)**	**0(0)**	**0(0)**	**0(0)**	**0(0)**
g13	3.8486×10^{-1}(0)	**0(0)**	9.7145×10^{-17}(0)	3.8486×10^{-1}(0)	0.6800(0)	**0(0)**	3.8486×10^{-1}	**0(0)**	1.1200×10^{-8}(0)	4.1898×10^{-11}(0)	0.9118(0)	0.3870(0)
g14	8.5123×10^{-12}(0)	**0(0)**	2.1316×10^{-14}(0)	6.3148×10^{-8}(0)	2.1316×10^{-14}(0)	**0(0)**	-3.9961×10^{-4}	**0(0)**	3.2912×10^{-3}(0)	1.4793×10^{-5}(0)	0.0174(0)	1.6397(0)
g15	6.0822×10^{-11}(0)	**0(0)**	**0(0)**	6.0936×10^{-11}(0)	**0(0)**	**0(0)**	**0(0)**	**0(0)**	1.0000×10^{-10}(0)	6.0822×10^{-11}(0)	**0(0)**	0.0009(0)
g16	6.5214×10^{-11}(0)	**0(0)**	4.4409×10^{-15}(0)	6.5216×10^{-11}(0)	5.1070×10^{-15}(0)	**0(0)**	5.3291×10^{-15}	**0(0)**	1.0000×10^{-10}(0)	6.5214×10^{-11}(0)	**0(0)**	**0(0)**
g17	7.4058×10^{1}(0)	7.4058×10^{1}(0)	1.8190×10^{-12}(0)	7.4052×10^{1}(0)	10.4896(0)	**0(0)**	7.4058×10^{1}	**0(0)**	13.9638(0)	7.4058×10^{1}(0)	**0(0)**	91.2351(0)
g18	1.5561×10^{-11}(0)	**0(0)**	3.3307×10^{-16}(0)	4.6362×10^{-11}(0)	4.4408×10^{-16}(0)	**0(0)**	4.4409×10^{-16}	**0(0)**	4.0000×10^{-10}(0)	1.5561×10^{-11}(0)	0.0001(0)	**0(0)**
g19	4.6370×10^{-11}(0)	**0(0)**	5.2162×10^{-8}(0)	5.2669×10^{-9}(0)	4.2632×10^{-14}(0)	0.387033(0)	3.5527×10^{-14}	**0(0)**	2.6302×10^{-2}(0)	1.3868×10^{-10}(0)	1.4028(0)	1.7022(0)
g20	-2.4674×10^{-2}(8)	-7.3330×10^{-2}(17)	-2.4674×10^{-2}(8)	1.3503×10^{1}(20)	0.1082(2)	0.103314(20)	1.0151×10^{1}	0.0675(12)	3.2600×10^{-2}(8)	2.3757×10^{-1}(20)	6.2057(8)	14.8045(9)
g21	-2.8371×10^{-10}(0)	3.5911×10^{-6}(0)	-2.8422×10^{-14}(0)	6.5523×10^{-8}(0)	-2.8421×10^{-14}(0)	**0(0)**	1.4211×10^{-13}	**0(0)**	81.3460(0)	2.5785×10^{-8}(0)	0.0633(0)	77.3185(0)
g22	1.0336×10^{4}(15)	1.2200×10^{2} (0)	1.2332×10^{1}(0)	9.7885×10^{3}(19)	8033.6537(8)	9210.082460(18)	8.7919×10^{3}	9888.6409(15)	14,198.8059(19)	4.6907×10^{1}(0)	10,565.5111(3)	8834.7836(3)
g23	3.0005×10^{2}(0)	1.0267×10^{-8} (0)	**0(0)**	1.0569×10^{1}(0)	2.2737×10^{-13}(0)	**0(0)**	**0(0)**	**0(0)**	130.5043(0)	3.9790×10^{-13}(0)	8.5726(0)	357.7081(0)
g24	4.6736×10^{-12}(0)	**0(0)**	5.7732×10^{-14}(0)	4.7269×10^{-12}(0)	5.5067×10^{-14}(0)	**0(0)**	7.1054×10^{-14}	**0(0)**	**0(0)**	4.6372×10^{-12}(0)	**0(0)**	**0(0)**

Table 9. Comparison of standard deviation values of JADE-ECHT, SaDE-ECHT and CEC'2006 algorithms achieved after 500,000 FEs. The bold numbers indicate the better results.

Prob	DE	DMS-PSO	ϵ DE	GDE	jDE-2	MDE	MPDE	PCX	PESO+	SaDE	JADE-ECHT	SaDE-ECHT
g01	6.4146×10^{-15}	0	0	0	0	0	0	0	0	0	0.4000	0
g02	1.0015×10^{-18}	4.6953×10^{-3}	1.7523×10^{-8}	7.5112×10^{-3}	3.0488×10^{-3}	0	2.7802×10^{-3}	0	7.1385×10^{-3}	4.9786×10^{-3}	0.0091	0.0254
g03	5.2098×10^{-14}	0	2.9582×10^{-31}	1.9833×10^{-1}	1.0140×10^{-1}	0	8.3681×10^{-2}	0	1.5396×10^{-6}	3.4743×10^{-5}	0.1036	0.0343
g04	3.9644×10^{-13}	0	0	0	0	0	0	0	0	1.8550×10^{-12}	0	0
g05	0	0	0	1.6854×10^{2}	2.4193	0	3.6380×10^{-13}	0	0	1.8190×10^{-13}	0	101.2496
g06	0	0	0	0	0	0	0	0	0	0	0	0
g07	6.4146×10^{-15}	0	2.1831×10^{-15}	3.8948×10^{-7}	1.7405×10^{-15}	0	3.0215×10^{-15}	0	2.2678×10^{-5}	1.4993×10^{-5}	0.0726	0.0850
g08	1.0015×10^{-18}	0	1.2326×10^{-32}	0	1.2580×10^{-32}	0	0	0	0	3.8426×10^{-18}	0	0
g09	5.2098×10^{-14}	0	0	4.9555×10^{-14}	4.2538×10^{-14}	0	0	0	0	7.9851×10^{-14}	0.0192	0.0002
g10	3.9644×10^{-13}	8.8141×10^{-9}	4.2426×10^{-13}	0	9.1279×10^{-8}	0	3.0165×10^{-13}	0	5.8279×10^{-2}	1.5244×10^{-6}	105.4501	3.0352
g11	0	0	0	0	4.5540×10^{-4}	0	7.0638×10^{-5}	0	0	0	0	0
g12	0	0	0	0	0	0	0	0	0	0	0	0
g13	0	1.8204×10^{-6}	0	3.8650×10^{-1}	2.2376×10^{-1}	0	2.8719×10^{-1}	0	6.3801×10^{-6}	2.7832×10^{-7}	0.1750	0.2828
g14	3.4809×10^{-15}	0	1.3924×10^{-15}	3.8556×10^{-3}	3.4809×10^{-15}	0	7.9441×10^{-15}	0	2.8553×10^{-2}	6.4986×10^{-5}	2.2940	0.6956
g15	0	0	0	9.6094×10^{-1}	2.2020×10^{-2}	0	4.3027×10^{-6}	0	0	0	0	0.6971
g16	2.2092×10^{-16}	0	1.5777×10^{-30}	1.7764×10^{-16}	0	0	0	0	0	0	0	0
g17	3.0234×10^{1}	0	1.2117×10^{-27}	8.2114×10^{1}	3.8319×10^{1}	0	3.4111×10^{1}	0	42.286	1.6168×10^{1}	24.5623	50.4589
g18	4.8817×10^{-17}	0	2.1756×10^{-17}	6.4619×10^{-1}	3.6822×10^{-17}	0	4.1541×10^{-17}	0	3.8184×10^{-2}	5.2898×10^{-2}	0.0041	0
g19	1.2568×10^{-5}	0	1.2568×10^{-5}	4.5735×10^{-5}	1.0531×10^{-13}	0.847517	3.5527×10^{-14}	0	1.6158×10^{-1}	7.2976×10^{-10}	1.0136	1.9699
g20	4.2362×10^{-2}	6.9516×10^{-3}	4.2362×10^{-2}	1.9580×10^{0}	1.1510×10^{-2}	0.021688	2.8984×10^{0}	0.0219	$\mathbf{4.9951 \times 10^{-3}}$	1.0638×10^{-1}	3.5087	4.8304
g21	6.5489×10^{1}	2.0784×10^{-6}	3.3417×10^{-14}	8.6788×10^{1}	3.6266×10^{1}	0	6.2358×10^{1}	0	67.019	1.5058×10^{-3}	63.9006	25.3368
g22	5.7875×10^{3}	2.7703×10^{1}	$\mathbf{1.5690 \times 10^{1}}$	5.9865×10^{3}	5.1748×10^{3}	4808.800969	4.8022×10^{3}	4421.5326	4963.7	3.0415×10^{1}	6162.3243	4998.2886
g23	8.8097×10^{-6}	4.5997×10^{-4}	1.1139×10^{-14}	1.9119×10^{2}	5.9983×10^{1}	0	5.1159×10^{-14}	0	87.634	3.4116×10^{-4}	198.2664	87.4764
g24	8.8525×10^{-20}	0	2.5244×10^{-29}	0	1.9323×10^{-29}	0	0	0	0	0	0	0

5. Conclusions and Future Work

This paper employed ECHT in the frameworks of two self-adaptive variants of DE, JADE and SaDE. Thus, constrained versions of the two algorithms, denoted by JADE-ECHT and SaDE-ECHT were developed. The proposed algorithms JADE-ECHT and SaDE-ECHT were tested and compared on CEC'06 benchmark test suit. The experimental results show that the SR of JADE-ECHT on most of the tested problems is better than SaDE-ECHT, while SaDE-ECHT surpasses JADE-ECHT in terms of FR. Both algorithms, like other algorithms in the literature, failed to solve problems g20 and g22 due to the hard nature of these problems. In the future, we intend to design ECHT of some other CHTs, embed it then in DE and swarm based algorithms to develop constrained evolutionary algorithms and finally test these newly developed algorithms on some real-world and engineering optimization problems. In addition to that we are going to use [34] for multipath routing protocols and for video streaming systems [35] in order to get the advantages of these plus the benefits of the proposed work would be very beneficent and demanded.

Author Contributions: Conceptualization, H.J. and M.A.J.; methodology, H.K., M.A.J. and W. K.M.; software, R.A.K., N.T. and H.S.; validation, H.U.K. and M.S.; formal analysis, H.J., M.A.J., R.A.K. and H.S.; investigation, H.J., M.A.J. and M.S.; resources, N.T. and H.S.; writing—original draft preparation, H.J. and M.A.J.; writing—review and editing, H.U.K. and M.S.; supervision, M.A.J. and W.K.M.; project administration, N.T. and H.S.; funding acquisition, N.T. and H.S.

Funding: The authors would like to thank King Khalid University of Saudi Arabia for supporting this research under the grant number R.G.P.2/7/38.

Conflicts of Interest: The authors declare no conflict of interest.

References

1. Storn, R.; Price, K. Differential evolution—A simple and efficient heuristic for global optimization over continuous spaces. *J. Glob. Optim.* **1997**, *11*, 341–359. [CrossRef]
2. Li, Z.; Shang, Z.; Liang, J.J.; Niu, B. An improved differential evolution for constrained optimization with dynamic constraint-handling mechanism. In Proceedings of the 2012 IEEE Congress on Evolutionary Computation (CEC), Brisbane, Australia, 10–15 June 2012; pp. 1–6.
3. Elsayed, S.M.; Sarker, R.A.; Essam, D.L. An improved self-adaptive differential evolution algorithm for optimization problems. *IEEE Trans. Ind. Inform.* **2013**, *9*, 89–99. [CrossRef]
4. Li, G.; Lin, Q.; Cui, L.; Du, Z.; Liang, Z.; Chen, J.; Lu, N.; Ming, Z. A novel hybrid differential evolution algorithm with modified CoDE and JADE. *Appl. Soft Comput.* **2016**, *47*, 577–599. [CrossRef]
5. Ali, M.; Kajee-Bagdadi, Z. A local exploration-based differential evolution algorithm for constrained global optimization. *Appl. Math. Comput.* **2009**, *208*, 31–48. [CrossRef]
6. Ameca-Alducin, M.Y.; Mezura-Montes, E.; Cruz-Ramírez, N. Dynamic differential evolution with combined variants and a repair method to solve dynamic constrained optimization problems: an empirical study. *Soft Comput.* **2018**, *22*, 541–570. [CrossRef]
7. Shah, T.; JAN, M.; Mashwani, W.K.; Wazir, H. Adaptive Differential Evolution for Constrained Optimization Problems. *Sci. Int.* **2016**, *28*, 2313–2320.
8. Wazir, H.; Jan, M.; Mashwani, W.; Shah, T. A penalty function based differential evolution algorithm for constrained optimization. *Nucleus* **2016**, *53*, 155–166.
9. Jan, M.A.; Khanum, R.A.; Tairan, N.M.; Mashwani, W.K. Performance of a Constrained Version of MOEA/D on CTP-series Test Instances. *Int. J. Adv. Comput. Sci. Appl.* **2016**, *7*, 496–505.
10. Brest, J.; Zumer, V.; Maucec, M.S. Self-adaptive differential evolution algorithm in constrained real-parameter optimization. In Proceedings of the IEEE Congress on Evolutionary Computation, Vancouver, BC, Canada, 16–21 July 2006; pp. 215–222.
11. Caraffini, F.; Kononova, A.V.; Corne, D. Infeasibility and structural bias in Differential Evolution. *arXiv* **2019**, arXiv:1901.06153.
12. Caraffini, F.; Kononova, A.V. Structural Bias in Differential Evolution: a preliminary study. *AIP Conf. Proc.* **2019**, *2070*, 020005.

13. Yaman, A.; Iacca, G.; Caraffini, F. A comparison of three differential evolution strategies in terms of early convergence with different population sizes. *AIP Conf. Proc.* **2019**, *2070*, 020002.

14. Iacca, G.; Neri, F.; Caraffini, F.; Suganthan, P.N. A differential evolution framework with ensemble of parameters and strategies and pool of local search algorithms. In Proceedings of the European Conference on the Applications of Evolutionary Computation, Granada, Spain, 23–25 April 2014; pp. 615–626.

15. Zhang, J.; Sanderson, A.C. JADE: Adaptive differential evolution with optional external archive. *IEEE Trans. Evol. Comput.* **2009**, *13*, 945–958. [CrossRef]

16. Caraffini, F.; Neri, F. A study on rotation invariance in differential evolution. *Swarm Evol. Comput.* **2018**. [CrossRef]

17. Caraffini, F.; Neri, F. Rotation invariance and rotated problems: An experimental study on differential evolution. In Proceedings of the International Conference on the Applications of Evolutionary Computation, Parma, Italy, 4–6 April 2018; pp. 597–614.

18. Liu, C.; Wang, G.; Xie, Q.; Zhang, Y. Vibration sensor-based bearing fault diagnosis using ellipsoid-ARTMAP and differential evolution algorithms. *Sensors* **2014**, *14*, 10598–10618. [CrossRef] [PubMed]

19. Datta, R.; Deb, K.; Kim, J.H. CHIP: Constraint Handling with Individual Penalty approach using a hybrid evolutionary algorithm. *Neural Comput. Appl.* **2018**, 1–17. [CrossRef]

20. Shakibayifar, M.; Hassannayebi, E.; Mirzahossein, H.; Taghikhah, F.; Jafarpur, A. An intelligent simulation platform for train traffic control under disturbance. *Int. J. Model. Simul.* **2019**, *39*, 135–156. [CrossRef]

21. Cheraitia, M.; Haddadi, S.; Salhi, A. Hybridizing plant propagation and local search for uncapacitated exam scheduling problems. *Int. J. Serv. Oper. Manag.* **2017**. [CrossRef]

22. Wang, G.G.; Tan, Y. Improving metaheuristic algorithms with information feedback models. *IEEE Trans. Cybern.* **2017**, *49*, 542–555. [CrossRef]

23. Fister, I.; Fister, I., Jr. *Adaptation and Hybridization in Computational Intelligence*; Springer: Berlin, Germany, 2015; Volume 18.

24. Wang, H.; Yi, J.H. An improved optimization method based on krill herd and artificial bee colony with information exchange. *Memet. Comput.* **2018**, *10*, 177–198. [CrossRef]

25. Coit, D.W.; Smith, A.E.; Tate, D.M. Adaptive penalty methods for genetic optimization of constrained combinatorial problems. *INFORMS J. Comput.* **1996**, *8*, 173–182. [CrossRef]

26. Michalewicz, Z.; Schoenauer, M. Evolutionary algorithms for constrained parameter optimization problems. *Evol. Comput.* **1996**, *4*, 1–32. [CrossRef]

27. Wolpert, D.H.; Macready, W.G. No free lunch theorems for optimization. *IEEE Trans. Evol. Comput.* **1997**, *1*, 67–82. [CrossRef]

28. Mallipeddi, R.; Suganthan, P.N. Ensemble of constraint handling techniques. *IEEE Trans. Evol. Comput.* **2010**, *14*, 561–579. [CrossRef]

29. Mallipeddi, R.; Suganthan, P.N. Differential evolution with ensemble of constraint handling techniques for solving CEC 2010 benchmark problems. In Proceedings of the 2010 IEEE Congress on Evolutionary Computation (CEC), Barcelona, Spain, 18–23 July 2010; pp. 1–8.

30. Liang, J.; Runarsson, T.P.; Mezura-Montes, E.; Clerc, M.; Suganthan, P.N.; Coello, C.C.; Deb, K. Problem definitions and evaluation criteria for the CEC 2006 special session on constrained real-parameter optimization. *J. Appl. Mech.* **2006**, *41*, 8–31.

31. Deb, K. An efficient constraint handling method for genetic algorithms. *Comput. Methods Appl. Mech. Eng.* **2000**, *186*, 311–338. [CrossRef]

32. Takahama, T.; Sakai, S. Constrained optimization by the ε constrained differential evolution with an archive and gradient-based mutation. In Proceedings of the 2010 IEEE Congress on Evolutionary Computation (CEC), Barcelona, Spain, 18–23 July 2010; pp. 1–9.

33. Runarsson, T.P.; Yao, X. Stochastic ranking for constrained evolutionary optimization. *IEEE Trans. Evol. Comput.* **2000**, *4*, 284–294. [CrossRef]

34. Iqbal, Z.; Khan, S.; Mehmood, A.; Lloret, J.; Alrajeh, N.A. Adaptive cross-layer multipath routing protocol for mobile ad hoc networks. *J. Sens.* **2016**, *2016*, 5486437. [CrossRef]
35. Taha, M.; Garcia, L.; Jimenez, J.M.; Lloret, J. SDN-based throughput allocation in wireless networks for heterogeneous adaptive video streaming applications. In Proceedings of the 2017 13th International Wireless Communications and Mobile Computing Conference (IWCMC), Valencia, Spain, 26–30 June 2017; pp. 963–968.

 *mathematics* 

Article

Memes Evolution in a Memetic Variant of Particle Swarm Optimization

Umberto Bartoccini [1], Arturo Carpi [2], Valentina Poggioni [2] and Valentino Santucci [1,*

[1] Department of Humanities and Social Sciences, University for Foreigners of Perugia, 06123 Perugia, Italy;
     umberto.bartoccini@unistrapg.it
[2] Department of Mathematics and Computer Science, University of Perugia, 1-06121 Perugia, Italy;
     arturo.carpi@unipg.it (A.C.); valentina.poggioni@unipg.it (V.P.)
*   Correspondence: valentino.santucci@unistrapg.it

Received: 1 April 2019; Accepted: 7 May 2019; Published: 11 May 2019

Abstract: In this work, a coevolving memetic particle swarm optimization (CoMPSO) algorithm is presented. CoMPSO introduces the memetic evolution of local search operators in particle swarm optimization (PSO) continuous/discrete hybrid search spaces. The proposed solution allows one to overcome the rigidity of uniform local search strategies when applied to PSO. The key contribution is that memes provides each particle of a PSO scheme with the ability to adapt its exploration dynamics to the local characteristics of the search space landscape. The objective is obtained by an original hybrid continuous/discrete meme representation and a probabilistic co-evolving PSO scheme for discrete, continuous, or hybrid spaces. The coevolving memetic PSO evolves both the solutions and their associated memes, i.e. the local search operators. The proposed CoMPSO approach has been experimented on a standard suite of numerical optimization benchmark problems. Preliminary experimental results show that CoMPSO is competitive with respect to standard PSO and other memetic PSO schemes in literature, and its a promising starting point for further research in adaptive PSO local search operators.

Keywords: memetic particle swarm optimization; adaptive local search operator; co-evolution; particle swarm optimization; PSO

---

1. Introduction

Memetic algorithms (MAs) are a class of hybrid evolutionary algorithms (EAs) and have been widely applied to many complex optimization problems [1], such as scheduling problems [2–4], combinatorial optimization problems [5–7], multi-objective optimization problems [8,9], and multimodal optimization problems [10,11]. Population-based evolutionary approaches, such as genetic algorithms (GAs) or particle swarm optimization (PSO) [12], are able to detect in a fast way the main regions of attraction, while their local search abilities represent a major drawback [13] from the point of view of solution accuracy and of the convergence behaviour, especially when applied to multimodal problems. MAs [1,14] have recently received attention as effective meta-heuristics to improve generic EA schemes by combining these latter with local search (LS) procedures [13,15]. In MAs, EA operations are employed for global rough exploration, and LS operators are used to execute further exploitation of single EA individuals. This allows one to enhance the EA's capability to solve complicated problems. In the context of PSO [12], the LS procedure can be regarded as an iterative local move of the particle. The scheme has been successfully applied to improve different meta-heuristics such as simulated annealing [16], GAs [17], and PSO [18]. A memetic algorithm that hybridizes PSO with LS, called MPSO, is proposed in [11] for locating multiple global and local optima in the fitness landscape of multimodal optimization problems (MMOPs). In this work, the local search is designed by means of two different operators, cognition-based local search (CBLS) and random walk with direction exploitation (RWDE), but no adaptation technique of the local search operators is employed.

Although many GA-based MAs and other EA approaches employing meme evolution have been proposed in the literature [13,19], the PSO memetic algorithms proposed so far fail to employ meme evolution, probably because of the difficulty of designing the descriptors of the local search operators, which are usually statically tailored to the problem, or the class of problems, at hand [7,8,10,11,20]. Indeed, they exhibit, at most, only some form of temporary adaptation that is not preserved or transmitted through the generations as meta-Lamarkian memes [21].

Since meme evolution is able to provide adaptivity of the local search operators, and many research results have shown that the advantages of the introduction of a local search in EAs, namely in PSO, are generalized, we have set as a research objective the investigation of the possibility of introducing memetic evolution of LS operators in a PSO context. The idea was to design a framework where memes are not selected in a preliminary evolutionary phase, but where they evolve and adapt during the search.

The research resulted in the coevolving memetic particle swarm optimization (CoMPSO) algorithm presented in this work. CoMPSO is an algorithm for numerical optimization in a hybrid sea and allows one to evolve local search operators in the form of memes associated with the PSO particles exploring a continuous and/or discrete hybrid search space. A special feature of local search operators, defined by the memes, is that they are applied in a probabilistic way. A meme is applied to the personal best position of the associated particle with a given probability and every given number of iterations. The basic idea is that memes are evolved themselves by means of the PSO-MEME evolution scheme, a classical PSO scheme modified to manage meme vectors composed of hybrid continuous/discrete components. Indeed, the PSO-MEME scheme distinguishes the evolution of continuous and discrete components where a probabilistic technique is applied.

The rest of the paper is structured as follows. Some general issues concerning MAs and PSO are recalled in Section 2.1, while the coevolving memetic PSO scheme is introduced in Section 2.2 by describing the proposed co-evolving particle-meme scheme and a probabilistic technique for PSO evolution of discrete integer components. Experimental results and comparisons with classical PSO and state-of-the-art static memetic PSO are presented and discussed in Section 3. Finally, in Section 4, a discussion of the advantages of the proposed approach and a description of future lines of research conclude the paper.

2. Co-Evolution of Memes in PSO: Models and Methods

In this section, we introduce the architecture of CoMPSO by first recalling some concepts of memetic algorithms and PSO, respectively in Sections 2.1.1 and 2.1.2. A detailed description of the CoMPSO scheme is then provided in Section 2.2, focusing on the algorithmic aspects introduced to guarantee diversity and on the representation for memes in the discrete, continuous, and hybrid space search, and finally describing and explaining the motivations for meme co-evolution strategies.

2.1. Memetic Algorithms and Particle Swarm Optimization

2.1.1. Memetic Algorithms

MAs, first proposed by Moscato [14,22], represent a recent growing area of research in evolutionary computation. The MA meta-heuristics essentially combine an EA with local search techniques that can be seen as a learning procedure that makes individuals capable of performing local refinements. MAs take inspiration from the Dawkins notion of "meme" [23] representing a unit of cultural evolution that can exhibit refinement.

MAs are usually distinguished into three main classes. First generation MAs [22] are characterized by a single fixed meme; i.e., a given local search operator is applied to candidate solutions of the main EA according to different selection and application criteria (see, for example, [20]). The criteria used in the selection of candidates and the frequency of the local search application represent the parameters of MAs. Second generation MAs, also called "meta-Lamarckian MAs" [21,24], reflect the principle of memetic transmission and selection; i.e., they are characterized by a pool of given local search operators (memes) that compete, based on their past merits, to be selected for future local refinements. The choice of the best memes can be based on different criteria, such as the absolute fitness of the candidate solutions associated with the meme or the fitness improvement due to past applications. Finally, the most recent generation of MA schemes are the co-evolving MAs [10,15,25,26], which introduce the idea of meme evolution. In this case, the pool of memes is not known a priori, but the memes are represented and evolved by means of EA methods similar to those applied for candidate solutions. The CoMPSO scheme proposed in this paper falls in this latter class.

2.1.2. Particle Swarm Optimization

Particle swarm optimization (PSO), originally introduced in [18], is a meta-heuristic approach to continuous optimization problems that is inspired by the collective behavior observed in flocks of birds. Further variants of PSO have been proposed in [12,27–33].

In PSO, a swarm of artificial entities—the so called *particles*—navigates the search space, aiming at optimizing a given objective/fitness function $f : \Theta \to \mathbb{R}$, where $\Theta \subseteq \mathbb{R}^d$ is the feasible region of the space. The set $P = \{p_1, \ldots, p_n\}$, composed of n particles, is endowed with a neighborhood structure; i.e., a set $N_i \subseteq P$ of neighbors is defined for each particle p_i. Every particle has a position in the search space, which is iteratively evolved using its own search experience and the experience of its neighbors. Therefore, PSO adopts both cognitive and social strategies [34] in order to focus the search on the most promising areas.

At every iteration $t \in \mathbb{N}$, any particle p_i is composed of the following d-dimensional vectors: the *position* in the search space $x_{i,t}$, the *velocity* $v_{i,t}$, the *personal best* position $b_{i,t}$ visited so far by the particle, and the *neighborhood best position* $g_{i,t}$ visited so far among the particles in N_i.

Both positions and velocities of the PSO particles are randomly initialized. Then they are iteratively updated until a stop criterion is met (e.g., the allowed budget of fitness evaluations has been consumed), according to the following *move equations* (first studied in [18]):

$$v_{i,t+1} = \omega v_{i,t} + \varphi_1 r_{1,t}(b_{i,t} - x_{i,t}) + \varphi_2 r_{2,t}(g_{i,t} - x_{i,t}), \tag{1}$$

$$x_{i,t+1} = x_{i,t} + v_{i,t+1}. \tag{2}$$

In Equation (1), the weight ω is the inertial coefficient, φ_1 and φ_2 are the acceleration factors, while the two random vectors $r_{1,t}, r_{2,t} \in \mathbb{R}^d$ are uniformly distributed in $[0, 1]^d$. The first term of Equation (1)—the *inertia* or *momentum*—is a memory of the trajectory followed by the particle so far. Its aim is to prevent drastic changes in the search performed by the particle. The second term is the *cognitive component* and represents the tendency of the particle to return to the best position ever visited by itself. Finally, the third term—the *social component*—models the contribution of the neighbors in the particle's trajectory.

After every position update, the fitness of $x_{i,t}$ is evaluated. If it is better than those of the previous personal and social best positions, these are updated accordingly.

Usually a fully connected topology is adopted as a neighborhood graph among the particles, so only one global social best g_t is needed instead of n local copies of it. It should be noted that sometimes the particles can exceed the bounds of the feasible search space. A common practice to address this issue, adopted also in our work, is to restart the out-of-bounds components of a particle in a position randomly chosen between the old position and the exceeded bound [35,36].

2.2. Coevolving Memetic PSO

Coevolving memetic PSO (CoMPSO) is a hybrid algorithm that combines PSO with the evolution of local search memetic operators that automatically adapt to the objective function landscape.

CoMPSO evolution can be seen as the combination of two co-evolving populations: the particles and the memes. The PSO particles represent the candidate solutions and evolve by means of the usual PSO laws (see Section 2.1.2). The memes represent the local search operators applicable to particles and are evolved by a modified PSO scheme that employs an innovative technique designed to deal with the discrete domains present in the meme representation. Furthermore, another difference with respect to the standard PSO scheme is the introduction of a diversity control mechanism that prevents the premature convergence of the particle population to a local optimum. Other example of the hybridization of PSO can be found in the area of pattern recognition of strings [37–40].

In the following, the general CoMPSO scheme (Section 2.2.1), the diversity control mechanism of particles evolution (Section 2.2.2), the meme representation (Section 2.2.3), meme evolution (Section 2.2.4), and the novel technique for managing discrete components in meme PSO (Section 2.2.5) are introduced.

2.2.1. The CoMPSO Scheme

In the CoMPSO general scheme, a population of n particles, i.e., $P = \{p_1, \ldots, p_n\}$, navigates the search space following PSO dynamics while trying to find the optimum of the given fitness function f. At each iteration, a local search is possibly applied to a subset of particles in order to improve their fitness. This local search phase is realized by a meme population, i.e., $M = \{m_1, \ldots, m_n\}$, of local search operators. Each m_i is associated with particle p_i and is evolved using a PSO-like approach on the meme space.

Particles in P are encoded using the usual d-dimensional vectors of the PSO scheme (see Section 2.1.2), i.e., the position vector $x_{i,t}^{(p)}$, the velocity vector $v_{i,t}^{(p)}$, and the personal best vector $b_{i,t}^{(p)}$, other than the common global best $g_t^{(p)}$. In this work, a complete neighborhood graph has been adopted. Similarly, a generic meme $m_j \in M$ is encoded by the hybrid vectors $x_{j,t}^{(m)}$, $v_{j,t}^{(m)}$, $b_{j,t}^{(m)}$, and $g_t^{(m)}$, i.e., vectors that combine both discrete and continuous components as described in Section 2.2.5.

The general CoMPSO scheme is described by the pseudo-code reported in Algorithm 1.

The populations of particles P and memes M are randomly initialized in their respective feasible regions. During main-loop iterations, each particle p_i is evolved following the scheme described in Section 2.1.2. *LOCAL_SEARCH* activates the memetic part of CoMPSO. The local search operators, defined by the memes in M, are applied in a probabilistic way: m_i is applied to the personal best position of particle p_i with a probability γ at every ϕ iteration, and the local search application is denoted by $m_i(p_i)$ or equivalently by $x_i^{(m)}(b_i^{(p)})$. Furthermore, for every iteration, the local search is also applied to the global best particle in position $g_t^{(p)}$. In this case, the global best particle and its respective meme are indicated by p_g and m_g. Before being applied, the memes are evolved by means of the *PSO_MEME* evolution scheme described in Section 2.2.4. If the meme application leads to a fitness improvement, the new candidate solution obtained replaces both the personal best that the particle current position. Finally, the global best is updated accordingly.

Algorithm 1 CoMPSO Pseudo-Code.

1: **procedure** CoMPSO
2:    $t \leftarrow 0$
3:    $P \leftarrow$ INITIALIZE_PARTICLES()
4:    $f(P) \leftarrow$ EVALUATE_FITNESS()
5:    $M \leftarrow$ INITIALIZE_MEMES()
6:    **while** termination criterion is not met **do**
7:       $t \leftarrow t + 1$
8:       **if** DIV_CONTROL() **then**
9:          DIV_RESTORE()
10:       **end if**
11:       **for all** $p_i \in P$ **do**
12:          $v_{i,t}^{(p)} \leftarrow$ UPDATE_VELOCITY$(v_{i,t-1}^{(p)}, x_{i,t-1}^{(p)}, b_{i,t-1}^{(p)}, g_{t-1}^{(p)})$
13:          $x_{i,t}^{(p)} \leftarrow$ UPDATE_POSITION$(x_{i,t-1}^{(p)}, v_{i,t}^{(p)})$
14:          $f(x_{i,t}^{(p)}) \leftarrow$ EVALUATE_FITNESS$(x_{i,t}^{(p)})$
15:          $b_{i,t}^{(p)} \leftarrow$ UPDATE_PERSONAL_BEST$(b_{i,t-1}^{(p)}, x_{i,t}^{(p)})$
16:       **end for**
17:       $g_t^{(p)} \leftarrow$ UPDATE_GLOBAL_BEST()
18:       **if** LOCAL_SEARCH(γ, ϕ) **then**
19:          **for all** $p_i \in P$ **do**
20:             PSO_MEME(m_i)
21:             $m_i(p_i)$
22:          **end for**
23:       **end if**
24:       PSO_MEME(m_g)
25:       $m_g(p_g)$
26:       $g_t^{(p)} \leftarrow$ UPDATE_GLOBAL_BEST()
27:    **end while**
28:    **return** getBestSolution
29: **end procedure**

2.2.2. Diversity Control

A diversity control mechanism has been introduced in order to avoid stagnation or premature convergence to local optima. When premature convergence is detected, a subset of particles whose positions will be reinitialized is selected. The method consists of two main components: a *diversity measure δ* and a *diversity restore mechanism* invoked when the population diversity becomes too low according to measure δ.

The diversity measure is computed on the fitness values of the particle population according to

$$\delta(P_t) \leftarrow \mathrm{std}(f(P_t)) \tag{3}$$

where $\mathrm{std}(f(P_t))$ represents the standard deviation of the fitness values of particle population P at time t. It should be noted that, although genotypic distances between particles seem to be in principle more appropriate than fitness values, it has been experimentally observed that the use of these latter indicators provides a good diversity measure with the additional property of a more efficient computation (this is due to the fact that the computation is performed on $\mathbb{R}$ and not on $\mathbb{R}^d$).

The diversity $\delta(P_0)$ of the initial random population P_0 is employed to compute the threshold value $\tau = 0.2 \cdot \delta(P_0)$ used in the routine *DIV_CONTROL* in order to recognize population convergence or stagnation. This threshold value is computed basing on the unbiased particle population at time 0 since it is the only one generated in a pure random way.

Therefore, at each iteration, in the case that $\delta(P_t) < \tau$, the *DIV_RESTORE* procedure is invoked. This routine randomly restarts the positions of the worst fit 50% of the population. Only the current positions of particles are restarted, while velocities and personal best values are kept unchanged.

2.2.3. Meme Representation

The local search operators adopted in CoMPSO are a generalization of the random walk (RW) technique [41–43]. RW is an iterative and stochastic local optimization method. Let y_t be the candidate (local) minimum at the t-th iteration. Then the new value y_{t+1} is computed according to

$$
y_{t+1} \leftarrow \begin{cases} y_t + w_t z_t & \text{if } f(x_t + w_t z_t) < f(x_t) \\ y_t & \text{otherwise} \end{cases} \tag{4}
$$

where w_t is a scalar step-length, and z_t is a unit-length random vector. The step length w_t is initialized to a given value w_0 and is halved at the end of every iteration in the case that the fitness has not been improved. The process is repeated until a given number q of iterations is reached.

In our generalization, other than the previously described parameters w_0 and q, two other parameters are introduced: the ramification factor b and the number of chosen successors k (with $k \leq b$). Briefly, the idea is to expand RW in a (b-k-bounded) breadth search conversely from the pure depth-first style of the original RW. Initially, k copies of the starting (or seed) point of RW are generated. Then b new points are generated from the current k ones following Rule (4). Therefore, from these b intermediate points, the k-fittest ones are chosen as next-iteration current points, and the process is iterated until the deepness parameter q is reached.

Using this local search scheme, a generic meme is represented by means of the following four parameters: a real value w_0 and the three integers b, k, and q. Furthermore, for each parameter, a range of admissible values has been experimentally established: $w_0 \in [0.5, 4]$, $b \in [1, 8] \cap \mathbb{N}$, $k \in [1, b] \cap \mathbb{N}$, and $q \in [4, 16] \cap \mathbb{N}$.

2.2.4. Meme Evolution

As introduced in the previous section, a meme is represented in the hybrid (continuous/discrete) meme space by the 4-ple $m = (w_0, b, k, q)$. Meme PSO evolution proceeds asynchronously with respect to particle evolution, this is due to the fact that memes are evolved only before they are applied and according to probability γ and frequency ϕ described above.

Memes, like particles, at each iteration t, are characterized by a position $x_{i,t}^{(m)}$, i.e., the representation of the meme in the meme space, a velocity $v_{i,t}^{(m)}$ (only for the continuous component), a personal best position $b_{i,t}^{(m)}$, and a global best meme $g_t^{(m)}$. Two alternative functions have been considered as meme fitness $f^{(m)}$: (1) the "absolute" fitness f of the particle p_i associated to meme m_i, and (2) the fitness improvement Δf realized with the application of the meme m_i to particle p_i. Some preliminary experiments have shown that the two approaches do not significantly differ.

Memes are evolved by a classical PSO scheme modified to manage the meme vectors composed by hybrid continuous/discrete components. The *PSO_MEME* scheme distinguishes the evolution of continuous components (according to Update Rules (1) and (2)) and that of discrete components where a probabilistic technique, presented in the next subsection, is applied. For the sake of completeness, *PSO_MEME* pseudo-code is reported in Algorithm 2.

Algorithm 2 Meme Evolution Pseudo-Code

1: **procedure** PSO_MEME(m_i)
2:     **for** each continuous dimension j **do**
3:         $v_{i,j}^{(m)} \leftarrow$ UPDATE_CONTINUOUS_VELOCITY($v_{i,j}^{(m)}, x_{i,j}^{(m)}, b_{i,j}^{(m)}, g_j^{(m)}$)
4:         $x_{i,j}^{(m)} \leftarrow$ UPDATE_CONTINUOUS_POSITION($x_{i,j}^{(m)}, v_{i,j}^{(m)}$)
5:     **end for**
6:     **for** each discrete dimension j **do**
7:         $x_{i,j}^{(m)} \leftarrow$ UPDATE_DISCRETE_POSITION($x_{i,j}^{(m)}, b_{i,j}^{(m)}, g_j^{(m)}$)
8:     **end for**
9:     $f^{(m)}(m_i) \leftarrow$ EVALUATE_MEME_FITNESS(m_i, p_i)
10:     $b_i^{(m)} \leftarrow$ UPDATE_MEME_PERSONAL_BEST($b_i^{(m)}, x_i^{(m)}$)
11:     $g^{(m)} \leftarrow$ UPDATE_MEME_GLOBAL_BEST($g^{(m)}, x_i^{(m)}$)
12: **end procedure**

2.2.5. Meme PSO for Discrete Domains

The meme discrete components are managed using a probabilistic technique that, by exploiting the total order of the integer domains here considered, simulates the classical PSO dynamics for continuous domains.

For each meme and for each discrete component domain $D_j = \{d_{j,1}, \ldots, d_{j,r}\}$, an appropriate probability distribution P_{D_j} over the values of D_j is built, then the new position x_j for component j is computed according to a randomized roulette wheel tournament among values in D_j and by using probabilities P_{D_j}.

As described in the following, the probability distribution is set in a way that simulates the properties of the classical continuous PSO [44].

Probability Distribution on Discrete Components Let $d_x, d_b, d_g \in D_j$ represents the values: the current meme position, the meme personal best position, and the global best meme position for component j of the meme encoding, then the distribution D_j is defined as a mixture of four discrete probability distributions:

1.   a uniform distribution that assigns probability $1/|D_j|$ (where $|D_j|$ is the cardinality of D_j) to every value $d_i \in D_j$,
2.   a triangular distribution centered in d_x,
3.   a triangular distribution centered in d_b, and
4.   a triangular distribution centered in d_g.

Each triangular distribution is determined by the center value d_k, an *amplification* factor $\alpha > 1$ and a *width* $2 \cdot \lambda + 1$. The probability of d_k is amplified by the factor α, i.e., the previous probability $P_{D_j}(d_k)$ is multiplied by α, while the $2 \cdot \lambda$ values of domain D_j located around d_k, i.e., $d_{k-\lambda}, \ldots, d_{k-1}$ and $d_{k+1}, \ldots, d_{k+\lambda}$, are amplified by a *smoothing* factor $\beta = \frac{\alpha}{\lambda+1} \cdot |\lambda + 1 - s|$, where s is the distance of the value from the center d_k. Moreover, it should be noted that the probabilities of each triangular distribution are normalized in order to sum up to 1.

The amplification factor α are set by using the PSO parameters; that is $1 + \omega$ for the center d_x, $1 + \varphi_1$ for d_b, and $1 + \varphi_2$ for d_g.

Discrete position update. The new position of a discrete component x_j is computed by a random roulette wheel tournament that uses the probability distribution above.

The technique for discrete components based on probability distribution values can be interpreted as initially considering all the values as equally probable, except for d_x, d_b, and d_g, whereas the amplification factors based on classical PSO parameters ω, φ_1, φ_2 confer a greater probability. Moreover, the smoothness factors, βs, also amplify the probabilities of centers neighbors, i.e., values close to the centers.

In other words, amplification and smoothing factors implement, for the discrete components, the probabilistic counterparts of the typical behavior of PSO velocity, i.e., the tendency to remain in the current position (inertia) and the tendency to move toward a personal and global best (cognitive and social dynamics). It should be noted that a similar idea is present in [45].

3. Experiments

The performances of CoMPSO have been evaluated on the set of five benchmark functions reported in Table 1. Those functions differ from each other for the properties of modality, symmetry around the optimum, and regularity.

In each run of CoMPSO, termination conditions have been defined for convergence and maximum computational resources. An execution is regarded convergent if $f(x) - f(x^{opt}) < \epsilon$. On the other hand, the execution has been considered terminated unsuccessfully if the number of function evaluations (NFES) exceeds the allowed cap of 100,000. The dimensionalities, feasible ranges, and ϵ values, used for each benchmark, are also reported in Table 1.

Table 1. Benchmark functions.

f	Dim.	Range	ϵ
$f_1(x) = \sum_{i=1}^{d} x_i^2$	30	$[-100, 100]^{30}$	10^{-2}
$f_2(x) = \sum_{i=1}^{d} x_i^2/4000 - \prod_{i=1}^{d} \cos(x_i/\sqrt{i}) + 1$	30	$[-600, 600]^{30}$	10^{-1}
$f_3(x) = 0.5 + \left\{ \sin^2(\sqrt{x_1^2 + x_2^2}) - 0.5 \right\} / \left\{ (1 + 0.001(x_1^2 + x_2^2))^2 \right\}$	2	$[-100, 100]^2$	10^{-5}
$f_4(x) = -20\exp\left\{ -0.2\sqrt{\frac{1}{d}\sum_{i=1}^{d} x_i^2} \right\} - \exp\left\{ \frac{1}{d}\sum_{i=1}^{d}\cos(2\pi x_i) \right\} + 20 + e$	30	$[-32, 32]^{30}$	10^{-3}
$f_5(x) = \sum_{i=1}^{d} \begin{cases} 0.15 \cdot (z_i - 0.05\mathrm{sgn}(z_i)^2) \cdot h_i & \text{if } |x_i - z_i| \leq 0.05 \\ h_i \cdot x_i^2 & \text{otherwise} \end{cases}$	4	$[-1000, 1000]^4$	10^{-7}

Following the setup of [20], three different swarm sizes have been considered, namely 15, 30, and 60. The other parameters were set by using the PSO standard values suggested in [46], i.e., $\omega = 0.7298$, $\varphi_1 = \varphi_2 = 1.49618$. The domain ranges defining the meme space were $b, k \in [1, 8] \cap \mathbb{N}$, $q \in [1, 16] \cap \mathbb{N}$, and $w \in [0.5, 4]$, while an amplification width $\lambda = 4$ was used for the evolution of the memes' discrete features. Furthermore, meme application probability and frequency were set to $\gamma = 0.2$ and $\phi = 5$.

For each swarm size configuration, a series of 50 executions was held in order to generate more confidence in the statistical results. In each execution series, the success rate SR (i.e., the number of convergent executions above the total number of executions), the average NFES of all convergent executions C, and the quality measure $Q_m = C_{avg}/SR$ introduced in [47] are recorded.

All these previously described performance indexes are reported in Table 2 for CoMPSO, classical PSO (CPSO), and static memetic PSO (SMPSO), i.e., PSO endowed with an RW local search operator but without meme evolution [20], which is the only comparable memetic PSO algorithm in the literature. The best quality measure in every line of Table 2 is provided in bold.

Table 2. Experimental results.

f	n	CoMPSO			CPSO			SMPSO		
		SR	*C*	Q_m	*SR*	*C*	Q_m	*SR*	*C*	Q_m
	15	1.00	14,324	**14,324**	0.00	-	-	0.00	-	-
f_1	30	1.00	12,226	12,226	1.00	11,235	**11,235**	0.50	37,409	74,818
	60	1.00	14,254	14,254	1.00	12,680	**12,680**	1.00	19,395	19,395
	15	0.00	-	-	0.00	-	-	0.00	-	-
f_2	30	0.96	12,306	**12,819**	0.50	8655	17,310	0.58	12,807	22,081
	60	1.00	15,497	15,497	1.00	11,480	**11,480**	0.70	15,785	22,550
	15	1.00	14,697	14,697	0.28	5475	19,553	0.83	6671	**8037**
f_3	30	1.00	24,433	24,433	0.64	10,935	**17,086**	1.00	27,247	27,247
	60	1.00	15,672	**15,672**	0.70	29,580	42,257	1.00	22,993	22,993
	15	1.00	45,184	**45,184**	0.00	-	-	1.00	57,385	57,385
f_4	30	1.00	41,077	**41,077**	0.00	-	-	1.00	43,673	43,673
	60	1.00	47,610	47,610	0.06	49,350	822,500	1.00	38,137	**38,137**
	15	1.00	2348	2348	1.00	1642	**1642**	0.98	2660	2714
f_5	30	1.00	3279	3279	1.00	2560	**2560**	1.00	3938	3938
	60	1.00	5206	5206	1.00	4220	**4220**	1.00	5285	5285

These results clearly show that the CoMPSO approach greatly improves the success rate of CPSO and SMPSO. In particular, it must be noted that CoMPSO converges almost everywhere, and it has a remarkable worst case convergence probability of 96%. On the other hand, CPSO, in some cases, fails to converge at all when a small swarm size is employed. Finally, the CoMPSO convergence speed is comparable to that of the SMPSO, although, as expected, in the more simple cases, i.e., f_1 and f_5, the convergence speed of CoMPSO is worse than that of CPSO, which is likely due to the NFES overhead introduced by local search operators.

Figures 1a,b and 2a–c plot the CoMPSO convergence graph with respect to the different swarm sizes adopted. These graphs show that CoMPSO appears quite monotonic with respect to swarm size and that a low number of particles seems to be generally preferable.

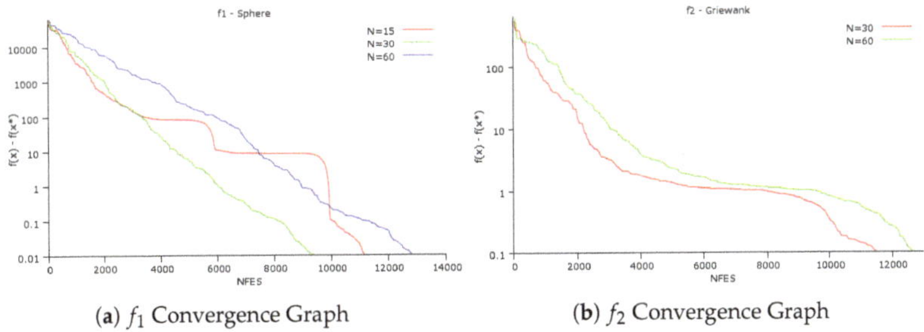

(**a**) f_1 Convergence Graph   (**b**) f_2 Convergence Graph

Figure 1. (**a**,**b**) CoMPSO convergence graphs for f_1 and f_2.

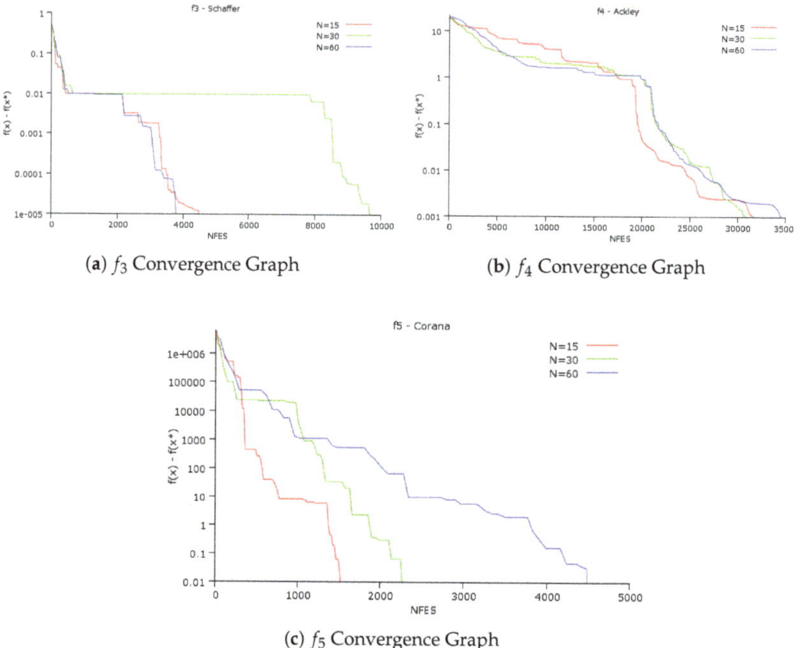

(**a**) f_3 Convergence Graph

(**b**) f_4 Convergence Graph

(**c**) f_5 Convergence Graph

Figure 2. (**a**–**c**) CoMPSO convergence graphs for f_3, f_4, and f_5.

Finally, a measure of meme convergence is shown in Figure 3, which shows the evolution of meme standard deviation (meme STD) on benchmark f_4. Meme convergence is fast in the early stage and, as expected, remains fairly constant, with only small adaptations, during the rest of the computation. The meme convergence curve, together with the quality measures and the success rates, show the effectiveness of CoMPSO and its ability to adapt its local refinement behavior to the landscape of the problem at hand.

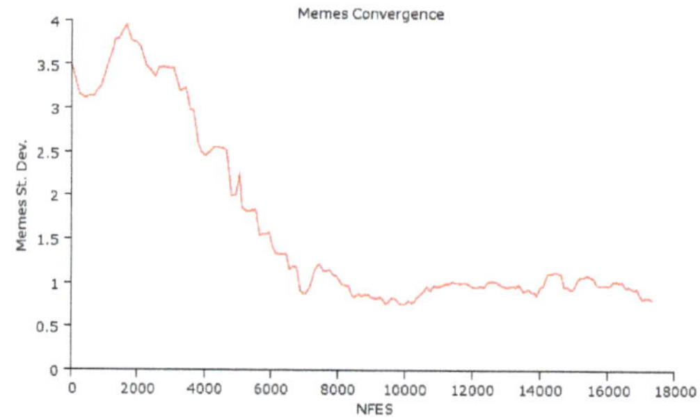

Figure 3. Memes Convergence Graph.

4. Discussion

In this paper, a coevolving memetic PSO (CoMPSO), characterized by two co-evolving populations of particles and memes, has been presented. The main contribution of this work is represented by the meme evolution technique that allows one to enhance the effectiveness of the PSO approach. Memetic algorithms (MAs) have recently received great attention as effective meta-heuristics to improve general evolutionary algorithm (EA) schemes by combining EAs with local search procedures and have demonstrated to be a very effective method for performance improvement. However, to the best of our knowledge, this is the first work where a memetic PSO algorithm with meme co-evolution has been proposed. Since memes have been described using one real and three integer parameters, a probabilistic PSO evolution technique for the discrete components in the meme representation has been designed. This technique is inspired from our previous work [45] and preserves typical PSO behavior of cognitive, social, and momentum dynamics.

The algorithm has been tested on some standard benchmark problems, and the results presented here show that CoMPSO outperforms the success rates of both classical PSO and static memetic PSO [20], although its convergence speed is affected by the overhead due to the local search applications. The effectiveness of the method relies on the ability to dynamically adapt the local search operators, i.e., the memes, to the problem landscape at hand. While experiments have been held on a limited set of standard benchmarks, the goal of demonstrating the feasibility of the approach i.e., providing the adaptivity of memes during searches in PSO, and improving the previous results can be considered achieved.

Future and ongoing works have been designed on different perspectives: from an experimental point of view, we are currently holding systematic experiments on larger sets of benchmarks, and we are planning experiments with hybrid continuous/discrete problems [48]; from the theoretical model point of view, we are investigating different models and synchronization mechanisms for the meme operators, we are also currently designing a self-regulatory mechanism for the CoMPSO parameters and investigating its applications to different classes of problems, such as multiobjective and multimodal problems [49].

Author Contributions: The authors equally contributed to this work, thus they share first-author contribution.

Funding: The research described in this work has been partially supported by: the research grant "Fondi per i progetti di ricerca scientifica di Ateneo 2019" of the University for Foreigners of Perugia under the project "Algoritmi evolutivi per problemi di ottimizzazione e modelli di apprendimento automatico con applicazioni al Natural Language Processing".

Acknowledgments: We would like to thank the reviewers for their valuable comments.

Conflicts of Interest: The authors declare no conflict of interest.

References

1. Neri, F.; Cotta, C. Memetic algorithms and memetic computing optimization: A literature review. *Swarm Evol. Comput.* **2012**, *2*, 1–14. [CrossRef]
2. Akhshabi, M.; Tavakkoli-Moghaddam, R.; Rahnamay-Roodposhti, F. A hybrid particle swarm optimization algorithm for a no-wait flow shop scheduling problem with the total flow time. *Int. J. Adv. Manuf. Technol.* **2014**, *70*, 1181–1188. [CrossRef]
3. Wu, X.; Che, A. A memetic differential evolution algorithm for energy-efficient parallel machine scheduling. *Omega* **2019**, *82*, 155–165. [CrossRef]
4. Liu, B.; Wang, L.; Jin, Y. An effective PSO-based memetic algorithm for flow shop scheduling. *IEEE Trans. Syst. Man Cybern.* **2007**, *37*, 18–27. [CrossRef]
5. Huang, H.; Qin, H.; Hao, Z.; Lim, A. Example-based Learning Particle Swarm Optimization for Continuous Optimization. *Inf. Sci.* **2012**, *182*, 125–138. [CrossRef]
6. Tang, J.; Lim, M.; Ong, Y. Diversity-adaptive parallel memetic algorithm for solving large scale combinatorial optimization problems. *Soft Comput.* **2007**, *11*, 873–888. [CrossRef]
7. Beheshti, Z.; Shamsuddin, S.M.; Hasan, S. Memetic binary particle swarm optimization for discrete optimization problems. *Inf. Sci.* **2015**, *299*, 58–84. [CrossRef]
8. Li, X.; Wei, J.; Liu, Y. A Memetic Particle Swarm Optimization Algorithm to Solve Multi-objective Optimization Problems. In Proceedings of the 2017 13th International Conference on Computational Intelligence and Security (CIS), Hong Kong, China, 15–18 December 2017; pp. 44–48. [CrossRef]

9. Liu, D.; Tan, K.C.; Goh, C.; Ho, W.K. A Multiobjective Memetic Algorithm Based on Particle Swarm Optimization. *IEEE Trans. Syst. Man Cybern. Part B Cybern.* **2007**, *37*, 42–50. [CrossRef]

10. Cao, Y.; Zhang, H.; Li, W.; Zhou, M.; Zhang, Y.; Chaovalitwongse, W.A. Comprehensive Learning Particle Swarm Optimization Algorithm with Local Search for Multimodal Functions. *IEEE Trans. Evol. Comput.* **2018**. [CrossRef]

11. Wang, H.; Moon, I.; Yang, S.; Wang, D. A Memetic Particle Swarm Optimization Algorithm for Multimodal Optimization Problems. *Inf. Sci.* **2012**, *197*, 38–52. [CrossRef]

12. Sengupta, S.; Basak, S.; Peters, R.A. Particle Swarm Optimization: A Survey of Historical and Recent Developments with Hybridization Perspectives. *Mach. Learn. Knowl. Extr.* **2018**, *1*, 157–191. [CrossRef]

13. Hart, W.; Krasnogor, N.; Smith, J. *Recent Advances in Memetic Algorithms*; Springer: Berlin/Heidelberg, Germany, 2005; Volume 166.

14. Cotta, C.; Mathieson, L.; Moscato, P., Memetic Algorithms. In *Handbook of Heuristics*; Martí, R., Pardalos, P.M., Resende, M.G.C., Eds.; Springer International Publishing: Cham, Switzerland, 2018; pp. 607–638. [CrossRef]

15. Gupta, A.; Ong, Y.S. Canonical Memetic Algorithms. In *Memetic Computation: The Mainspring of Knowledge Transfer in a Data-Driven Optimization Era*; Springer International Publishing: Cham, Switzerland, 2019; pp. 17–26. [CrossRef]

16. Kirkpatrick, S.; Gelatt, C.; Vecchi, M. Optimization by simulated annealing. *Science* **1983**, *220*, 671. [CrossRef]

17. Michalewicz, Z. *Genetic Algorithms + Data Structures = Evolution Programs*; Springer: Berlin, Germany, 1992; Volume 19.

18. Kennedy, J.; Eberhart, R. Particle swarm optimization. In Proceedings of the 1995 IEEE International Conference on Neural Networks, Perth, Western Australia, 27 November–1 December 1995; Volume 4, pp. 1942–1948.

19. Eusuff, M.; Lansey, K.; Pasha, F. Shuffled frog-leaping algorithm: A memetic meta-heuristic for discrete optimization. *Eng. Optim.* **2006**, *38*, 129–154. [CrossRef]

20. Petalas, Y.; Parsopoulos, K.; Vrahatis, M. Memetic particle swarm optimization. *Ann. Oper. Res.* **2007**, *156*, 99–127. [CrossRef]

21. Ong, Y.; Keane, A. Meta-Lamarckian learning in memetic algorithms. *IEEE Trans. Evol. Comput.* **2004**, *8*, 99–110. [CrossRef]

22. Moscato, P. On evolution, search, optimization, genetic algorithms and martial arts: Towards memetic algorithms. In *Caltech Concurrent Computation Program, C3P Rep.*; California Institute of Technology: Pasadena, CA, USA, 1989; Volume 826, p. 1989.

23. Dawkins, R. *The Selfish Gene*; Oxford University Press: New York, NY, USA, 2006.

24. Carpi, A.; de Luca, A. Central Sturmian Words: Recent Developments. In *Developments in Language Theory 2005*; Lecture Notes in Computer Science; Springer: Berlin/Heidelberg, Germany, 2005; Volume 3572, pp. 36–56.

25. Smith, J. Coevolving memetic algorithms: A review and progress report. *IEEE Trans. Syst. Man Cybern. Part B Cybern.* **2007**, *37*, 6–17. [CrossRef]

26. Baioletti, M.; Milani, A.; Santucci, V. A New Precedence-Based Ant Colony Optimization for Permutation Problems. In *Simulated Evolution and Learning*; Springer International Publishing: Cham, Switzerland, 2017; pp. 960–971. [CrossRef]

27. Baioletti, M.; Milani, A.; Santucci, V. Algebraic Particle Swarm Optimization for the permutations search space. In Proceedings of the 2017 IEEE Congress on Evolutionary Computation, CEC 2017, Donostia, San Sebastián, Spain, 5–8 June 2017; pp. 1587–1594. [CrossRef]

28. Ahmed, A.; Sun, J. Bilayer Local Search Enhanced Particle Swarm Optimization for the Capacitated Vehicle Routing Problem. *Algorithms* **2018**, *11*, 31. [CrossRef]

29. Yu, X.; Estevez, C. Adaptive Multiswarm Comprehensive Learning Particle Swarm Optimization. *Information* **2018**, *9*, 173. [CrossRef]

30. Baioletti, M.; Milani, A.; Santucci, V. Automatic Algebraic Evolutionary Algorithms. In Proceedings of the International Workshop on Artificial Life and Evolutionary Computation (WIVACE 2017), Venice, Italy, 19–21 September 2017; pp. 271–283. [CrossRef]

31. Santucci, V.; Baioletti, M.; Milani, A. A Differential Evolution Algorithm for the Permutation Flowshop Scheduling Problem with Total Flow Time Criterion. In Proceedings of the 13th International Conference on Parallel Problem Solving from Nature—PPSN XIII, Ljubljana, Slovenia, 13–17 September 2014; pp. 161–170. [CrossRef]
32. Dai, H.; Chen, D.; Zheng, Z. Effects of Random Values for Particle Swarm Optimization Algorithm. *Algorithms* **2018**, *11*, 23. [CrossRef]
33. Letting, L.; Hamam, Y.; Abu-Mahfouz, A. Estimation of Water Demand in Water Distribution Systems Using Particle Swarm Optimization. *Water* **2017**, *9*, 593. [CrossRef]
34. Akarsh, S.; Kishor, A.; Niyogi, R.; Milani, A.; Mengoni, P. Social Cooperation in Autonomous Agents to Avoid the Tragedy of the Commons. *Int. J. Agric. Environ. Inf. Syst.* **2017**, *8*, 1–19. [CrossRef]
35. Oldewage, E.T.; Engelbrecht, A.P.; Cleghorn, C.W. Boundary Constraint Handling Techniques for Particle Swarm Optimization in High Dimensional Problem Spaces. In *Swarm Intelligence*; Dorigo, M., Birattari, M., Blum, C., Christensen, A.L., Reina, A., Trianni, V., Eds.; Springer International Publishing: Cham, Switzerland, 2018; pp. 333–341.
36. De Luca, A.; Carpi, A. Uniform words. *Adv. Appl. Math.* **2004**, *32*, 485–522. [CrossRef]
37. Hamed, H.N.A.; Kasabov, N.; Michlovský, Z.; Shamsuddin, S.M.H. String Pattern Recognition Using Evolving Spiking Neural Networks and Quantum Inspired Particle Swarm Optimization. In Proceedings of the 16th International Conference on Neural Information Processing, ICONIP 2009, Bangkok, Thailand, 1–5 December 2009; pp. 611–619. [CrossRef]
38. D'Alessandro, F.; Carpi, A. Independent sets of words and the synchronization problem. *Adv. Appl. Math.* **2013**, *50*, 339–355. [CrossRef]
39. De Luca, A.; Carpi, A. Harmonic and gold Sturmian words. *Eur. J. Comb.* **2004**, *25*, 685–705. [CrossRef]
40. Carpi, A. On the repetition threshold for large alphabets. In *Mathematical Foundations of Computer Science 2006*; Lecture Notes in Computer Science; Springer: Berlin/Heidelberg, Germany, 2006; Volume 4162, pp. 226–237.
41. Bordenave, C.; Caputo, P.; Salez, J. Random walk on sparse random digraphs. *Probab. Theory Relat. Fields* **2018**, *170*, 933–960. [CrossRef]
42. Weiss, G. *Aspects and Applications of the Random Walk (Random Materials & Processes S.)*; North-Holland: New York, NY, USA, 1994.
43. Mengoni, P.; Milani, A.; Li, Y. Clustering students interactions in eLearning systems for group elicitation. In *Computational Science and Its Applications—ICCSA 2018*; Lecture Notes in Computer Science; Springer: Cham, Switzerland, 2018; Volume 10962, pp. 398–413. [CrossRef]
44. Baioletti, M.; Milani, A.; Santucci, V. Learning Bayesian Networks with Algebraic Differential Evolution. In *Parallel Problem Solving from Nature—PPSN XV*; Springer International Publishing: Cham, Switzerland, 2018; pp. 436–448. [CrossRef]
45. Santucci, V.; Milani, A. Particle Swarm Optimization in the EDAs Framework. In *Soft Computing in Industrial Applications*; Springer: Berlin/Heidelberg, Germany, 2011; pp. 87–96. [CrossRef]
46. Clerc, M.; Kennedy, J. The particle swarm-explosion, stability, and convergence in a multidimensional complex space. *IEEE Trans. Evol. Comput.* **2002**, *6*, 58–73. [CrossRef]
47. Feoktistov, V. *Differential Evolution: In Search of Solutions*; Springer: New York, NY, USA, 2006; Volume 5.
48. Baioletti, M.; Milani, A.; Santucci, V. An Extension of Algebraic Differential Evolution for the Linear Ordering Problem with Cumulative Costs. In *Parallel Problem Solving from Nature—PPSN XIV*; Springer: Cham, Switzerland, 2016; pp. 123–133. [CrossRef]
49. Baioletti, M.; Milani, A.; Santucci, V. MOEA/DEP: An Algebraic Decomposition-Based Evolutionary Algorithm for the Multiobjective Permutation Flowshop Scheduling Problem. In *Evolutionary Computation in Combinatorial Optimization—EvoCOP 2018*; Springer International Publishing: Cham, Switzerland, 2018; pp. 132–145. [CrossRef]

Article

A Multi-Objective Particle Swarm Optimization Algorithm Based on Gaussian Mutation and an Improved Learning Strategy

Ying Sun [1,†] and Yuelin Gao [1,2,*,†]

1   School of Computer Science and Information Engineering, Hefei University of Technology, Hefei 230009, China; sunying@nun.edu.cn
2   Ningxia Province Key Laboratory of Intelligent Information and Data Processing, North Minzu University, Yinchuan 750021, China
*   Correspondence: 1993001@nun.edu.cn; Tel.: +86-951-206-6579
†   These authors contributed equally to this work.

Received: 8 January 2019; Accepted: 29 January 2019; Published: 4 February 2019

Abstract: Obtaining high convergence and uniform distributions remains a major challenge in most metaheuristic multi-objective optimization problems. In this article, a novel multi-objective particle swarm optimization (PSO) algorithm is proposed based on Gaussian mutation and an improved learning strategy. The approach adopts a Gaussian mutation strategy to improve the uniformity of external archives and current populations. To improve the global optimal solution, different learning strategies are proposed for non-dominated and dominated solutions. An indicator is presented to measure the distribution width of the non-dominated solution set, which is produced by various algorithms. Experiments were performed using eight benchmark test functions. The results illustrate that the multi-objective improved PSO algorithm (MOIPSO) yields better convergence and distributions than the other two algorithms, and the distance width indicator is reasonable and effective.

Keywords: multi-objective optimization problems; particle swarm optimization (PSO); Gaussian mutation; improved learning strategy

---

1. Introduction

Multi-objective optimization problems (MOPs) are very common in engineering and other areas of research, such as economics, finance, production scheduling, and aerospace engineering. It is very difficult to solve these problems because they usually involve several conflicting objectives. Generally, the optimal solution of MOPs is a set of optimal solutions (known as a Pareto optimal set), which differs from the solution of single-objective optimization (with only one optimal solution) [1]. Some classical optimization methods (weighted methods, goal programming methods, etc.) require the problem functions to be differentiable and are required to run multiple times with the hope of finding different solutions. In recent years, the multi-objective optimization evolutionary algorithm (MOEA) has become a popular method for solving MOPs, and it has garnered scholarly interest around the world [2]. Many representative MOEAs, such as multiple objective genetic algorithm (MOGA) [3], non-dominated sorting genetic algorithm II (NSGA-II) [4], strength pareto evolutionary algorithm (SPEA) [5] and pareto archived evolution strategy (PAES) [6], have been presented.

Over the past decade, the particle swarm optimization algorithm (PSO) has been used to solve MOPs, and a number of multi-objective PSO algorithms have been suggested. Some studies have shown that the modified PSO algorithm can effectively solve the MOPs and that the non-dominated solution set of the algorithm is much closer to the true Pareto optimal front. Coello and Pultdo [7] incorporated a special mutation operator that enriches the exploratory capabilities. Sierra and Coello [8] proposed the OMOPSO algorithm, which places the whole population into three subpopulations with the same size and uses different mutation operators within different subpopulations. This algorithm improves the exploration ability of the particles. Reddy and Kumar [9] proposed an elitist-mutation multi-objective PSO (EM-MOPSO) algorithm with a strategic mechanism that effectively explores the feasible search space and speeds up the search for the true Pareto optimal region. Leong and Yen [10] proposed a dynamic population multiple-swarm MOPSO algorithm that uses an adaptive local archive to improve the diversity within each swarm. Yen and Leong [11] also proposed a dynamic population multiple-swarm MOPSO algorithm in which the number of swarms is adaptively adjusted throughout the search process via the proposed dynamic swarm strategy. The strategy allocates an appropriate number of swarms to support convergence and diversity criteria among the swarms as required. Chen, Zou, and Wang [12] presented a multi-objective endocrine particle swarm optimization algorithm (MOEPSO) in which the hormone (RH), released by the endocrine system, is encoded as a particle swarm and is then supervised by the corresponding stimulating hormone. Lin et al. [13] introduced a novel MOPSO algorithm using multiple search strategies (MMOPSO), and a decomposition approach was used to transform MOPs into a set of aggregation problems. Then, each particle was accordingly assigned to optimize each aggregation problem. A multi-objective vortex particle swarm optimization (MOVPSO) method was proposed in [14] based on the emerging properties of a swarm to simulate motion with diversity control via collaborative mechanisms using linear and circular movements. The parallel cell coordinate system (PCCS) in self-adaptive MOPSO [15] is used to select global and personal bests, maintain archives and adjust flight parameters. Knight et al. [16] presented a spreading mechanism to promote diversity in MOPSO. Cheng et al. [17] presented a hybrid multi-objective particle swarm optimization that combines the canonical PSO search with a teaching–learning-based optimization (TLBO) algorithm to promote diversity and improve the search ability. Overall, for any improved MOPSO algorithm, the search ability is determined by the neighbouring topological structure, and the convergence rate depends on the dominance relationship.

In this paper, we introduce a new multi-objective PSO algorithm based on Gaussian mutation and an improved learning strategy to solve MOPs. The main new contributions of this work can be summarized as: (1) Gaussian mutation throw points strategy to improve the uniformity of external archives and current populations; (2) For MOPs, it is difficult to select the g_{best} value of velocity and update the formula. Unlike other MOPSOs, that often randomly select a solution from the external archive as the global optimal solution g_{best}, we present different learning strategies to update the individual positions of the non-dominated and dominated solutions; (3) To further measure the distribution width, the indicator DW is proposed.

The remainder of this paper is organized as follows. In Section 2, we describe the multi-objective optimization. Thereafter, in Section 3, we present a multi-objective improved PSO algorithm (MOIPSO). Section 4 outlines the MOIPSO algorithm. Test problems, performance measures, and the results are provided in Section 5, and the conclusions are presented in Section 6.

2. Description of Multi-Objective Optimization Problems

A general minimization problem of m objectives can be mathematically stated as follows:

Given $x = (x_1, x_2, \cdots, x_n) \in D, D \subset R^n$ where n is the dimension of decision variable space D. Additionally,

$$\begin{aligned} min \quad & y = f(x) = [f_1(x), f_2(x), \cdots, f_m(x)] \\ s.t. \quad & g_i(x) \leq 0, \quad i = 1, 2, \cdots, p \\ & h_j(x) = 0, \quad j = 1, 2, \cdots, q \end{aligned} \tag{1}$$

where $y = (f_1, f_2, \cdots, f_n) \in Y$ is the objective function vector, Y is the objective variable space, $g_i(x)$ is the i-th inequality constraint, and $h_j(x)$ is the j-th equality constraint.

Multiple objectives are included in MOPs, therefore, it is not possible to find a single solution that can optimize all objectives. Generally, improving one objective may cause the performance of the other objectives to decrease. Therefore, the conventional concept of single-objective optimality does not hold, and we must find a solution that is a compromize based on all objectives, i.e., the Pareto optimality. Based on the aforementioned reasons, some important definitions are given as follows for MOPs [18,19]:

Definition 1. *(Pareto dominance) The vector $x' = (x'_1, x'_2, \cdots, x'_n)$ dominates the vector $x = (x_1, x_2, \cdots, x_n)$ if and only if the next statement is verified. $\forall i = (1, 2, \cdots, n) : f_i(x') \leq f_i(x)$ and $\exists i \in (1, 2, \cdots, n) : f_i(x') < f_i(x)$, denoted as $x' \prec x$.*

Definition 2. *(Pareto optimality) A solution $x^* \in D$ is a Pareto optimal solution if there is not another $x \in D$ that satisfies $f(x) \prec f(x^*)$.*

Definition 3. *(Pareto optimal set) The Pareto optimal set is defined as the set of all Pareto optimal solutions.*

Definition 4. *(Pareto optimal front) The Pareto front consists of the values of the objectives corresponding to the solutions in the Pareto optimal set.*

3. An Introduction to the Multi-Objective Improved PSO

3.1. Main Aspects of the Standard PSO Algorithm

PSO was first presented by Kennedy and Eberhart in 1995 [20]. It is a random optimization algorithm based on swarm aptitude. The theory behind PSO comes from research on the behaviour of a bird swarm catching food. Compared with genetic algorithms, it has a simple construction, can be easily implemented, and has few adjustable parameters (Algorithm 1).

Let n be the dimension of the search space, $x_i = (x_{i1}, x_{i2}, \cdots, x_{in})$ be the current position of the i-th particle in the swarm, $p_{ibest} = (p_{ibest1}, p_{ibest2}, \cdots, p_{ibestn})$ be the best position of the i-th particle at that time, and $g_{best} = (g_{best1}, g_{best2}, \cdots, g_{bestn})$ be the best position that the whole swarm has visited. The rate of the velocity of the i-th particle is denoted as $v_i = (v_{i1}, v_{i2}, \cdots, v_{in})$.

Algorithm 1: Standard particle swarm optimization [20]

Step 1: Initialize a population of particles X_N, such that each particle has a random position vector x_i and a velocity vector v_i. Set parameters c_1 and c_2, the maximum number of generations T_{max}, and the generation number $T = 0$.

Step 2: Calculate the fitness of all the particles in $X_N(T)$.

Step 3: Renew the positions and velocities of particles based on the following equations:

$$v_{id}^{T+1} = wv_{id}^T + c_1 r_1 \left(p_{ibestd} - x_{id}^T\right) + c_2 r_2 \left(g_{bestd} - x_{id}^T\right) \tag{2}$$

$$x_{id}^{T+1} = x_{id}^T + v_{id}^{T+1}. \tag{3}$$

Step 4: Calculate the fitness of the particles and renew every optimal position and global optimal position of the particles.

Step 5: (Termination examination) If the termination criterion is satisfied, then output the global optimal position and the fitness value. Otherwise, let $T = T + 1$ and return to **Step 2**.

3.2. Main Aspects of the Multi-Objective Improved PSO Algorithm (MOIPSO)

3.2.1. Elitist Archive and Crowding Entropy

Since Zitzler introduced SPEA with an elitist reservation mechanism in 1999 [5], many new algorithms have adopted a similar elitist reservation mechanism. Namely, they provide an external archive to store all the non-dominated solutions that have been found. The elitist reservation mechanism is also adopted in this article. As the evolution progresses, the non-dominated solutions in the current archive may not be the non-dominated solutions in the entire evolutionary process, therefore, the archive must be updated. The easiest updating method compares each solution with the current archive at each generation, which allows for the input of better solutions into the archive. The specific archive update rules are as follows [21]: (I) If the new solution dominates one or more solutions of the external archive, the new solution enters the archive, and the dominated solutions are deleted from the archive; (II) If the new solution is dominated by one or more solutions in the external archive, then the new solution is rejected; (III) If the new solution and the solutions in the external archive are not dominated by each other, then the new solution is a non-dominated solution and enters the archive. However, because of the storage space and computational efficiency, the external archive is not infinite. When the archive reaches its maximum size, the largest crowding degree solution will be deleted.

For the crowding distance measure, we cite the crowding entropy in the literature [21]. This method combines the crowded distance and distribution entropy, and the method accurately measures the crowding degree of the solution.

Crowding entropy is defined as follows:

$$
\begin{aligned}
CE_i &= \sum_{j=1}^{m}(c_{ij}E_{ij})/(f_j^{\max} - f_j^{\min}) \\
&= -\sum_{j=1}^{m}[dl_{ij}\log_2(pl_{ij}) + du_{ij}\log_2(pu_{ij})]/(f_j^{\max} - f_j^{\min}),
\end{aligned}
\tag{4}
$$

where E_{ij} is the distribution entropy of the i-th solution to the j-th objective function. Specifically, E_{ij} is defined as $E_{ij} = -[pl_{ij}\log_2(pl_{ij}) + pu_{ij}\log_2(pu_{ij})]$, where $pl_{ij} = \frac{dl_{ij}}{c_{ij}}$, $pu_{ij} = \frac{du_{ij}}{c_{ij}}$, and $c_{ij} = dl_{ij} + du_{ij}$. Variables dl_{ij} and du_{ij} are the distances from the i-th solution to the lower and upper adjacent solutions for the j-th objective function, $f_j^{\max}$ and $f_j^{\min}$ are the maximum and minimum values of the j-th objective function, and m is the number of objective functions.

Thus, the smaller the crowding entropy, the more crowded the archive. For each objective function, the boundary solutions are assigned infinite crowding entropy values. All other intermediate solutions are assigned crowding entropy values according to Equation (4).

3.2.2. Gaussian Mutation Strategy

Gaussian mutation is a very popular way to improve the particle swarm optimization algorithm. Higashi et al. [22] integrate a Gaussian mutation used for GA into PSO, and leave a certain ambiguity in the transition to the next generation due to Gaussian mutation. This method is used to solve the single-objective optimization problem, and carries on each individual variation in the current population. For the multi-objective problems, Coelho et al. [23] used Gaussian mutation to update the velocity update formula, but they only replaced the uniform random number R with a Gaussian random number Gd in the velocity formula. Liang et al. [24] also introduced Gaussian mutation, which will have a certain probability to initialize the particle adjacent to the target particle. Meanwhile, this will randomly initialize the particles beyond the range to increase the utilization rate of particles. To further improve the performance of the solutions from the MOPs, this paper presents a new Gaussian mutation throw point strategy, which involves the throwing of points into external archives and the current population. The details of the strategy are as follows.

1. Throw points at sparse positions in the external archive to produce a thickened point set (TPS).

For MOPs, researchers hope that the non-dominated solution set is evenly distributed in the true Pareto front, but the solution sets of many methods yield uneven distributions. To increase the number of solutions at sparse positions and make the distribution of the solution set more uniform, we define the crowding degree of the solution in the external archive based on the crowding entropy and use throw points based on a Gaussian distribution of the largest crowding entropy solution (except the boundary solutions). Note that it is more important to find the boundary solutions for the MOPs, and the crowding entropy is infinite at the boundaries. Therefore, we must throw points at the boundary solutions.

The concrete operations are as follows.

Step 1: Identify a sparse solution based on the crowding entropy, as shown in Figure 1.

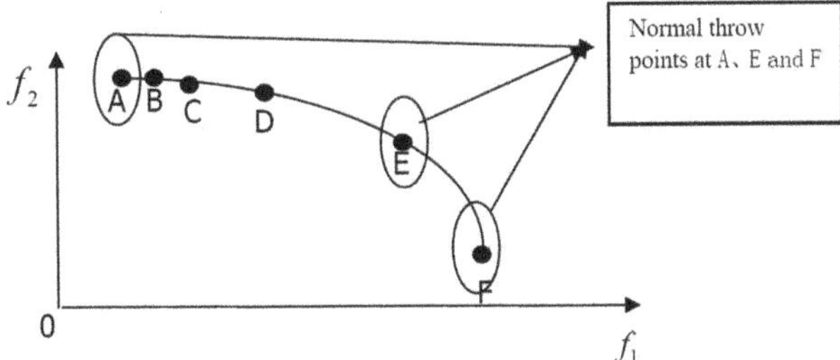

Figure 1. The selection process of the sparse solution.

Step 2: In the decision space, $A \longrightarrow x_A$, $E \longrightarrow x_E$, and $F \longrightarrow x_F$. We normally throw h points based on centres x_A, x_E, and x_F and variance σ. The random variable $Z \sim N(A, \sigma)$, and we add h random points to the TPS. It should be noted that the value of the variance σ is $\frac{1}{5}$ the width of each dimension. For example; $x = (x_1, x_2) \in [-10, 0] \times [0, 10] \Longrightarrow \sigma = \begin{pmatrix} 2 & 0 \\ 0 & 2 \end{pmatrix}; h = 5.$

2. Throw points into the current population to produce a Gaussian mutation points set (GMPS).

The algorithm chooses R solutions based on the prescribed probability in the current population, and normal throw points are established at R solutions (method based on (i): **Step 2**). Finally, we obtain a new GMPS.

3.2.3. Improved Learning Strategy

The standard PSO algorithm is used to solve the single-objective optimization problem, therefore, it is difficult to select the g_{best} value of velocity and update the formula for MOPs. The reason for this issue is that MOPs do not contain the global optimal solution. In many previous articles, researchers have randomly selected a solution from the external archive as the global optimal solution g_{best}, but this method lacks pertinence and cannot reflect the guidance of g_{best}. Therefore, this article presents a modified velocity formula, redefines the value g_{best} of Equation (2) and more efficiently applies g_{best} to solve MOPs.

(i) When x_i^T is not in the external archive, all solutions in the archive that dominate x_i^T can be regarded as global optimal solutions. Therefore, we provide a linear combination of these solutions.

Suppose that there are k archive solutions that dominate $x_i^T : a_1^T, a_2^T, \cdots, a_k^T$. A set of weights w_j is randomly generated, where $j = 1, 2, \cdots, k$ and $\sum_{j=1}^{k} w_j = 1$. Thus, g_{best} can be expressed as follows:

$$g_{\text{best}} = \sum_{j=1}^{k} w_j a_j^T.$$

Velocity is updated in the formula as follows:

$$v_{id}^{T+1} = w v_{id}^T + c_1 r_1 (p_{ibestd} - x_{id}^T) + c_2 r_2 (g_{bestd} - x_{id}^T). \tag{5}$$

(ii) When x_i^T is in the external archive, the concept of a global optimal solution is meaningless for x_i^T because the global optimal solution is a non-dominated solution. Therefore, all other solutions cannot be better than this solution, and g_{best} does not exist. Hence, the three parts of Equation (2) are unnecessary, and the velocity updating formula is as follows:

$$v_{id}^{T+1} = w v_{id}^T + c_1 r_1 (p_{ibestd} - x_{id}^T). \tag{6}$$

The position update formula still uses the original model: $x_{id}^{T+1} = x_{id}^T + v_{id}^{T+1}$.

3.2.4. Update External Archive

The updating process of the external archive is an important problem for MOPs. Researchers typically use the archive update rules to compare the current population and the old external archive and then generate a new external archive to further improve the performance of the external archive. This paper uses three sets to update the old external archive. The specific updating methods are shown in Figure 2.

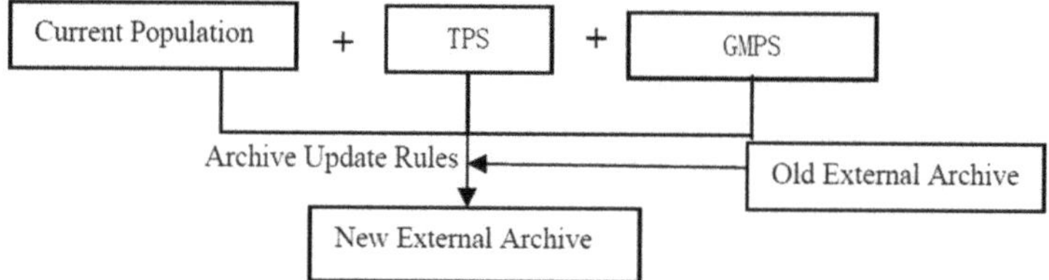

Figure 2. The updating process of the external archive.

3.2.5. Population Elitist Incremental Strategy

To increase the convergence rate of the population when the algorithm generates the offspring population, we consider the effect of not only the parent population but also the external archive. This paper proposes an elitist incremental strategy that increases the number of external archive solutions in the offspring population to form a new offspring population. The definition is as follows:

$$\begin{aligned}
&\text{A new offspring population} \\
&= \text{randomly selected}(N - L)\text{offspring solutions} \\
&\quad + \text{randomly selected } L \text{ external archive solutions,}
\end{aligned} \tag{7}$$

where N is the population size,

$$
L = \begin{cases}
T & T \le \left\lceil \frac{N}{2} \right\rceil \ \& \ |A| \ge T \\
|A| & T \le \left\lceil \frac{N}{2} \right\rceil \ \& \ |A| < T \\
\left\lceil \frac{N}{2} \right\rceil & T > \left\lceil \frac{N}{2} \right\rceil \ \& \ |A| \ge \left\lceil \frac{N}{2} \right\rceil \\
|A| & T > \left\lceil \frac{N}{2} \right\rceil \ \& \ |A| < \left\lceil \frac{N}{2} \right\rceil
\end{cases} \ ,
$$

T is the number of iterations, and $|A|$ is the external archive of size A at iteration T.

As a result, the population has strong exploration abilities and can find a wider range of non-dominated solutions at the early stage. The population will further explore the known non-dominated solutions and move closer to the true Pareto front in the later stage.

4. Overview of the MOIPSO Algorithm

As previously discussed, the MOIPSO algorithm can be summarized as follows.

Step 1: Randomly initialize the position and velocity of each particle within the search space.

(a) Set the following parameters; $c_1 = c_2$, v_{max}, v_{min}, w_{max}, w_{min}, $T = 0$, the Gaussian mutation probability p_r, the maximum generation number T_{max}, the population size N, and the maximum external archive size A_{max}.

(b) Randomly initialize the population and the velocity.

Step 2: Calculate the fitness of the particles in the initialized population, and initialize the optimal position p_{ibest} of the i-th particle.

Step 3: Initialize an update to the external archive A.

Step 4: For $T = 1$:

(a) Renew the velocities of particles based on Equations (5) and (6) and the position of particles based on Equation (7) to form the middle population;

(b) Select the particles based on the Gaussian mutation probability p_r and throw points using the Gaussian mutation strategy (ii) to produce a GMPS;

(c) Calculate the crowding entropy of the external archive solutions and throw points based on the Gaussian mutation strategy (i) to produce a TPS;

(d) Renew the external archive A based on Section 3.2.4. If $|A| > A_{max}$, then delete the most crowded particles according to the crowding entropy;

(f) Renew the middle population using the elitist incremental strategy, and form the new offspring population;

(g) Calculate the fitness of the particle in the offspring population;

(h) Renew the optimal position p_{ibest} of each particle;

(i) If the termination criterion is satisfied, then output the Pareto optimal solutions. Otherwise, let $T = T + 1$ and go to Step (a).

5. Methods and Simulation Experiments

5.1. Test Problems

To test the performance of MOIPSO, eight unconstrained optimization problems were used in the experiments. The SCH, KUR, and FON functions were suggested by Schaffer in 1985 [25], Kursawe in 1991 [26], and Fonseca in 1998 [27], respectively. The remainders are ZDT problems suggested by Zitzler et al. in 2000 [28]. The optimization problems are described in Table 1.

Table 1. The tested optimization problems.

Function	Objective Functions	D	Variable Bounds	Characteristics of the Pareto Front
SCH	$\begin{cases} f_1(x) = x^2 \\ f_2(x) = (x-2)^2 \end{cases}$	1	$x \in [-10^3, 10^3]$	Convex
FON	$\begin{cases} f_1(x) = 1 - \exp(-\sum_{i=1}^{3}(x_i - \frac{1}{\sqrt{3}})^2) \\ f_2(x) = 1 - \exp(-\sum_{i=1}^{3}(x_i + \frac{1}{\sqrt{3}})^2) \end{cases}$	3	$x_i \in [-4,4]$	Nonconvex
KUR	$\begin{cases} f_1(x) = \sum_{i=1}^{n-1}(-10exp(-0.2\sqrt{x_i^2 + x_{i+1}^2})) \\ f_2(x) = \sum_{i=1}^{n}(|x_i|^{0.8} + 5\sin(x_i^3)) \end{cases}$	3	$x_i \in [-5,5]$	Disconnect
ZDT1	$\begin{cases} f_1(x) = x_1 \\ f_2(x) = g(x)[1 - \sqrt{x_1/g(x)}] \\ g(x) = 1 + 9(\sum_{i=2}^{n} x_i)/(n-1) \end{cases}$	30	$x_i \in [0,1]$	Convex
ZDT2	$\begin{cases} f_1(x) = x_1 \\ f_2(x) = g(x)[1 - (x_1/g(x))^2] \\ g(x) = 1 + 9(\sum_{i=2}^{n} x_i)/(n-1) \end{cases}$	30	$x_i \in [0,1]$	Nonconvex
ZDT3	$\begin{cases} f_1(x) = x_1 \\ f_2(x) = g(x)(1 - \sqrt{x_1/g(x)} - x_1\sin(10\pi x_1)/g(x)) \\ g(x) = 1 + 9\sum_{i=2}^{n} x_i/(n-1) \end{cases}$	30	$x_i \in [0,1]$	Convex disconnect
ZDT4	$\begin{cases} f_1(x) = x_1 \\ f_2(x) = g(x)(1 - \sqrt{x_1/g(x)}) \\ g(x) = 1 + 10(n-1) + \sum_{i=2}^{n}[x_i^2 - 10\cos(4\pi x_i)] \end{cases}$	10	$x_1 \in [0,1]$ $x_i \in [-5,5]$ $i = 2, \cdots, n$	Nonconvex
ZDT6	$\begin{cases} f_1(x) = 1 - \exp(-4x_1)\sin^6(6\pi x_1) \\ f_2(x) = g(x)[1 - (f_1(x)/g(x))^2] \\ g(x) = 1 + 9[\sum_{i=2}^{n} x_i/(n-1)]^{0.25} \end{cases}$	10	$x_i \in [0,1]$	Nonconvex

5.2. Performance Measures

The standard performance measures of multi-objective evolutionary algorithms were used to evaluate the performance of the proposed algorithm. The performance measures are briefly described as follows.

5.2.1. Convergence Measure Indicator

Ideally, the iterative process of MOEA approaches the Pareto front, but in most cases, it is difficult to find the true Pareto front. The proximity of the approximate solutions to the Pareto optimal solutions is a main indicator.

The concept of generational distance was introduced by Van Veldhuizen [29] to measure the proximity of the approximate solutions to the Pareto optimal solutions. This indicator is defined as follows:

$$GD = \frac{\sqrt{\sum_{i=1}^{n} \text{dist}_i^2}}{n}, \tag{8}$$

where n is the number of non-dominated solutions. When the Pareto fronts of the objective function can be expressed in analytic form, dist_i is measured by the Euclidean distance (in objective space) between the i-th non-dominated solution and the nearest member of the Pareto optimal set. Otherwise, dist_i is measured by the Euclidean distance between the i-th non-dominated solution and the reference point set. It is clear that a smaller value of GD is better, and $GD = 0$ indicates that the non-dominated solution set is located in the true Pareto front.

5.2.2. Distribution Measure Indicator

Typically, we hope that non-dominated solutions are uniformly distributed in the true Pareto front. Two main factors used to measure the distribution are uniformity and width:

(1) Distribution uniformity (Δ)

The indicator Δ [4] is used to measure the uniformity and diversity of the non-dominated solution set. When calculating this indicator, we must sort the obtained non-dominated solutions based on the specified objective function values. This indicator is defined as follows:

$$\Delta = \frac{h_f + h_l + \sum\limits_{i=1}^{n-1} |h_i - \bar{h}|}{h_f + h_l + (n-1)\bar{h}},$$

(9)

where n is the number of non-dominated solutions, h_i is the Euclidean distance between neighbouring solutions in the non-dominated solution set, $\bar{h}$ is the mean of all h_i values, h_f and h_l are the Euclidean distances between the extreme solutions of the Pareto optimal solution set and the boundary solutions of the non-dominated solution set. When the non-dominated solution set is uniformly distributed in the true Pareto front, $h_f = 0$, $h_l = 0$, and all $h_i = \bar{h}$, therefore, $\Delta = 0$. A small value of Δ indicates better uniformity in the true Pareto front.

(2) Distribution width (DW)

Generally, it is favourable if the boundary solutions can be included in the non-dominated solution set. In other words, in these types of problems, researchers hope to find the boundary points of the true Pareto fronts.

The indicator $M_3^*(NP)$, which measures the distribution width, was proposed by Zitzler et al. [28]. The associated formula is as follows:

$$M_3^*(NP) = \sqrt{\sum_{i=1}^{M} \max\{\| p_i - q_i \|^*, p, q \in NP\}},$$

(10)

where NP is the non-dominated solution set and M is the dimension of the non-dominated solutions. Notably, M_3^* can measure the distribution width of the non-dominated solution set, but when the distribution range of the Pareto front is too large or the dimensions of the solutions are large, the value of M_3^* will be large. Therefore, it is difficult to compare and compile the data.

Based on the aforementioned method, the new indicator, DW, is provided. The concrete form of this indicator is as follows:

$$DW = \left| \frac{\prod\limits_{i=1}^{M} \max\{|p_i - q_i|, p, q \in NP\}}{\prod\limits_{i=1}^{M} \max\{|p_i' - q_i'|, p', q' \in RP\}} - 1 \right|,$$

(11)

where RP is the Pareto optimal set or the reference point set. Notably, a small value of DW reflects a better distribution uniformity for the non-dominated solution set.

5.3. Algorithm Comparison

To validate the MOIPSO algorithm, we compared it to NSGA-II [4] and MOPSO [7] based on the above three performance measures. The source codes of NSGA-II and MOPSO are available at http://delta.cs.cinvestav.mx/~ccoello/EMOO/ (matlab code).

The initial population size is 100 for MOIPSO, NSGA-II and MOPSO. The number of iterations directly affects the time complexity of the algorithm, and the convergence of the problem with different iterations is discussed [30]. It can be found that when the program reaches a stable state, the subsequent iterations can no longer improve the performance of the algorithm, but only increase the running time. Therefore, through a large number of numerical experiments, different iteration numbers were chosen based on the complexity of the problems in the three algorithms. The number of iterations completed was 60 for SCH, 100 for FON and KUR, 300 for ZDT1, ZDT2, and ZDT3, and 1000 for ZDT4 and ZDT6.

The range of the parameter has been discussed in some literature [31], so we took the commonly used parameter values for NSGAII and MOPSO. To make the test and its results more comparable, for MOIPSO, the same parameters as MOPSO took the same values, and other parameters were set after a large number of numerical experiments. In the MOPSO and MOIPSO algorithms, $c_1 = c_2 = 1.7$, while r_1 and r_1 are assigned random values between 0 and 1. In MOPSO, the inertia weight is $w = 0.7$, and the inertia weight damping ratio $w_{damp} = 1$. In MOIPSO, the inertia weight w adaptively decreased from $w_{max} = 0.9$ to $w_{min} = 0.4$ according to the following formula: $w = w_{max} - \frac{T(w_{max} - w_{min})}{T_{max}}$. In the NSGA-II [4] algorithm, the crossover probability is 0.8, and the mutation probability is 0.3. To evaluate the statistical performance, all the experiments are run 30 times. The best, worst, mean, and average deviations are shown in Tables 2–4, respectively.

The *GD* results are shown in Table 2. MOIPSO exhibits the best *GD* values for SCH, KUR, ZDT1 and ZDT4. MOPSO displays the best *GD* values for ZDT2, ZDT3, and ZDT6. FON, MOIPSO, and MOPSO exhibit similar results. For the worst *GD*, MOIPSO displays high stability and consistently yields the lowest worst *GD* value in all test problems. For the mean *GD*, MOIPSO produces the best mean values for all test problems except ZDT4, for which FON, MOIPSO, and MOPSO exhibit similar results. With respect to the standard deviation of *GD*, MOIPSO exhibits the best solution for KUR, ZDT1, ZDT2, and ZDT3. NSGA-II yields the best results for the other functions. Thus, MOIPSO produces better values of *GD* indicators than the other two algorithms in most test problems, and the results of MOIPSO are better than those of the other algorithms by 1∼2 orders of magnitude. This finding indicates that the resulting Pareto fronts obtained via MOIPSO are closer to the true Pareto fronts, and MOIPSO can effectively improve convergence.

Some information for Δ is shown in Table 3. For the SCH and ZDT1 functions, all the solutions of MOIPSO are better than those of the other algorithms. MOPSO exhibits the best, and the mean best, Δ value for KUR. MOIPSO has the minimal worst Δ and the best standard deviation. MOIPSO has the best, the mean best, and the minimal worst Δ values for ZDT2 and ZDT3, but the standard deviation of NSGA-II is the best. MOIPSO displays the best and the mean best Δ values for ZDT4. NSGA-II exhibits the minimal worst Δ, and MOPSO yields the best standard deviation. MOPSO provides the best Δ for ZDT6. MOIPSO has the minimal worst Δ and the mean best Δ, and the minimal standard deviation of NSGA-II is the best. Table 3 shows that MOIPSO provides the best mean solution for all seven functions. Therefore, the MOIPSO results are evenly distributed in the experiments, however, they are not shown for all the functions.

Table 4 shows the results of a new quality indicator—*DW*. All the solutions of MOIPSO are better than those of the other algorithms for SCH, KUR, and ZDT1. MOIPSO yields the three best indicators for the FON function, and MOPSO provides the best solution for *DW*. NSGA-II exhibits the best *DW* solution for ZDT2, and the other indicators of MOIPSO are the best. MOIPSO displays the best and the mean best *DW* for ZDT3, and NSGA-II yields the minimal worst *DW* and the best standard deviation. NSGA-II exhibits the best *DW* for ZDT4, and MOPSO displays the minimal worst *DW*. MOIPSO produces the mean best *DW* and the minimal standard deviation. For the ZDT6 function, NSGA-II exhibits the minimal standard deviation, and the other indicators of MOIPSO are the best. Similarly, the *DW* results in Table 4 show that MOIPSO is able to produce the best distribution of solutions in the Pareto optimal front for most of the test functions.

Table 2. Comparison of the results of multi-objective improved particle swarm optimization algorithm (MOIPSO) and different algorithms based on the indicator, GD.

Function	Statistic	MOPSO	NSGA-II	MOIPSO
SCH	Best	8.72×10^{-4}	8.37×10^{-4}	8.34×10^{-4}
	Worst	1.30×10^{-3}	1.10×10^{-3}	1.00×10^{-3}
	Mean	9.66×10^{-4}	9.56×10^{-4}	9.41×10^{-4}
	Std	7.89×10^{-5}	4.70×10^{-5}	4.76×10^{-5}
FON	Best	1.00×10^{-3}	1.10×10^{-3}	1.00×10^{-3}
	Worst	1.30×10^{-3}	1.30×10^{-3}	1.20×10^{-3}
	Mean	1.10×10^{-3}	1.20×10^{-3}	1.10×10^{-3}
	Std	6.22×10^{-5}	3.73×10^{-5}	6.08×10^{-5}
KUR	Best	3.10×10^{-3}	1.60×10^{-3}	1.40×10^{-3}
	Worst	3.09×10^{-2}	4.40×10^{-3}	3.80×10^{-3}
	Mean	5.80×10^{-3}	3.00×10^{-3}	2.50×10^{-3}
	Std	4.82×10^{-3}	6.43×10^{-4}	4.13×10^{-4}
ZDT1	Best	1.70×10^{-3}	1.96×10^{-2}	5.91×10^{-4}
	Worst	6.43×10^{-2}	5.49×10^{-2}	3.20×10^{-3}
	Mean	2.63×10^{-2}	3.03×10^{-2}	1.00×10^{-3}
	Std	1.81×10^{-2}	7.80×10^{-3}	6.30×10^{-4}
ZDT2	Best	5.28×10^{-5}	2.96×10^{-2}	3.51×10^{-4}
	Worst	1.55×10^{-1}	4.99×10^{-2}	3.10×10^{-3}
	Mean	4.19×10^{-2}	3.84×10^{-2}	7.52×10^{-4}
	Std	4.49×10^{-2}	5.40×10^{-3}	5.94×10^{-4}
ZDT3	Best	4.04×10^{-4}	2.19×10^{-2}	5.01×10^{-4}
	Worst	1.08×10^{-1}	4.74×10^{-2}	7.10×10^{-3}
	Mean	3.82×10^{-2}	3.30×10^{-2}	1.20×10^{-3}
	Std	3.27×10^{-2}	8.10×10^{-3}	1.50×10^{-3}
ZDT4	Best	1.10×10^{-1}	1.30×10^{-3}	3.23×10^{-4}
	Worst	2.01×10^{-1}	2.86×10^{-2}	1.61×10^{-2}
	Mean	1.40×10^{-1}	4.90×10^{-3}	9.90×10^{-3}
	Std	1.67×10^{-2}	6.20×10^{-3}	2.37×10^{-2}
ZDT6	Best	2.72×10^{-4}	1.19×10^{-1}	5.36×10^{-4}
	Worst	3.32×10^{-1}	1.65×10^{-1}	4.38×10^{-2}
	Mean	2.29×10^{-2}	1.46×10^{-1}	1.09×10^{-2}
	Std	5.96×10^{-2}	9.10×10^{-3}	1.13×10^{-2}

Table 3. Comparison of the results of MOIPSO and different algorithms based on Δ.

Function	Statistic	MOPSO	NSGA-II	MOIPSO
SCH	Best	4.68×10^{-1}	4.68×10^{-1}	4.67×10^{-1}
	Worst	4.80×10^{-1}	4.73×10^{-1}	4.70×10^{-1}
	Mean	4.68×10^{-1}	4.69×10^{-1}	4.67×10^{-1}
	Std	2.80×10^{-3}	1.70×10^{-3}	4.53×10^{-4}
FON	Best	4.82×10^{-1}	4.81×10^{-1}	4.84×10^{-1}
	Worst	4.94×10^{-1}	4.95×10^{-1}	4.85×10^{-1}
	Mean	4.86×10^{-1}	4.87×10^{-1}	4.85×10^{-1}
	Std	2.60×10^{-3}	3.90×10^{-3}	4.70×10^{-4}
KUR	Best	4.59×10^{-1}	4.64×10^{-1}	4.63×10^{-1}
	Worst	4.69×10^{-1}	4.69×10^{-1}	4.68×10^{-1}
	Mean	4.63×10^{-1}	4.67×10^{-1}	4.66×10^{-1}
	Std	2.10×10^{-3}	1.30×10^{-3}	1.10×10^{-3}
ZDT1	Best	4.08×10^{-2}	2.67×10^{-1}	1.13×10^{-2}
	Worst	3.54×10^{-1}	4.05×10^{-1}	1.08×10^{-1}
	Mean	2.47×10^{-1}	3.43×10^{-1}	4.28×10^{-2}
	Std	8.62×10^{-2}	3.71×10^{-2}	9.70×10^{-3}

Table 3. *Cont.*

Function	Statistic	MOPSO	NSGA-II	MOIPSO
ZDT2	Best	3.78×10^{-1}	4.09×10^{-1}	3.90×10^{-3}
	Worst	9.92×10^{-1}	4.90×10^{-1}	1.47×10^{-1}
	Mean	6.88×10^{-1}	4.59×10^{-1}	2.65×10^{-2}
	Std	1.54×10^{-1}	1.99×10^{-2}	3.17×10^{-2}
ZDT3	Best	4.19×10^{-1}	2.64×10^{-1}	2.50×10^{-3}
	Worst	7.06×10^{-1}	3.60×10^{-1}	1.94×10^{-1}
	Mean	5.58×10^{-1}	3.13×10^{-1}	3.08×10^{-2}
	Std	6.58×10^{-2}	2.94×10^{-2}	4.57×10^{-2}
ZDT4	Best	6.11×10^{-1}	1.33×10^{-2}	3.70×10^{-3}
	Worst	6.89×10^{-1}	3.50×10^{-1}	4.74×10^{-1}
	Mean	6.50×10^{-1}	8.96×10^{-2}	8.88×10^{-2}
	Std	1.95×10^{-2}	9.90×10^{-2}	1.50×10^{-1}
ZDT6	Best	1.07×10^{-2}	4.86×10^{-1}	1.11×10^{-2}
	Worst	4.52×10^{-1}	7.95×10^{-1}	4.08×10^{-1}
	Mean	2.43×10^{-1}	6.29×10^{-1}	2.28×10^{-1}
	Std	1.28×10^{-1}	6.92×10^{-2}	1.28×10^{-1}

Table 4. Comparison of the results of MOIPSO and different algorithms based on *DW*.

Function	Statistic	MOPSO	NSGA-II	MOIPSO
SCH	Best	1.00×10^{-3}	4.46×10^{-5}	4.44×10^{-5}
	Worst	6.43×10^{-2}	4.35×10^{-2}	1.80×10^{-2}
	Mean	1.43×10^{-2}	1.66×10^{-2}	2.70×10^{-3}
	Std	1.56×10^{-2}	1.26×10^{-2}	3.70×10^{-3}
FON	Best	2.62×10^{-4}	4.21×10^{-4}	1.90×10^{-3}
	Worst	3.36×10^{-2}	9.09×10^{-2}	8.00×10^{-3}
	Mean	9.60×10^{-3}	3.10×10^{-2}	5.50×10^{-3}
	Std	9.30×10^{-3}	2.19×10^{-2}	1.60×10^{-3}
KUR	Best	3.07×10^{-4}	5.50×10^{-3}	2.40×10^{-4}
	Worst	7.55×10^{-2}	7.36×10^{-2}	1.11×10^{-2}
	Mean	1.32×10^{-2}	3.71×10^{-2}	3.30×10^{-3}
	Std	1.82×10^{-2}	1.95×10^{-2}	2.40×10^{-3}
ZDT1	Best	7.80×10^{-3}	5.00×10^{-2}	3.20×10^{-3}
	Worst	1.33	1.29	2.26×10^{-2}
	Mean	5.23×10^{-1}	4.98×10^{-1}	1.17×10^{-2}
	Std	4.05×10^{-1}	2.83×10^{-1}	6.20×10^{-2}
ZDT2	Best	1.54×10^{-1}	1.60×10^{-3}	3.30×10^{-3}
	Worst	1.00	6.40×10^{-1}	2.99×10^{-1}
	Mean	7.72×10^{-1}	2.07×10^{-1}	4.16×10^{-2}
	Std	2.55×10^{-1}	1.49×10^{-1}	6.28×10^{-2}
ZDT3	Best	4.60×10^{-3}	3.49×10^{-2}	1.83×10^{-4}
	Worst	8.93×10^{-1}	5.54×10^{-1}	9.63×10^{-1}
	Mean	6.23×10^{-1}	2.59×10^{-1}	8.52×10^{-2}
	Std	2.37×10^{-1}	1.54×10^{-1}	2.37×10^{-1}
ZDT4	Best	1.92×10^{-1}	9.28×10^{-4}	6.50×10^{-3}
	Worst	1.00	1.64	1.68
	Mean	6.33×10^{-1}	2.21×10^{-1}	2.07×10^{-1}
	Std	1.95×10^{-1}	4.03×10^{-1}	1.55×10^{-1}
ZDT6	Best	1.42×10^{-2}	1.04×10^{-2}	2.20×10^{-4}
	Worst	6.87	1.17	1.09
	Mean	1.18	2.78×10^{-1}	1.86×10^{-1}
	Std	1.48	2.78×10^{-1}	8.77×10^{-1}

According to the above statistical analyses, MOIPSO successfully solves the SCH, ZDT1 and ZDT6 problems, as illustrated by Figures 3–5. Figure 3 shows that the three algorithms have similar convergence performances for the SCH function, however, MOIPSO exhibits better uniformity. Figures 4 and 5 illustrate the superior convergence and distribution results of MOIPSO compared with those of the other algorithms. Furthermore, by combining the data from Table 4 and the figures, we can see that when the boundary solutions do not exist in non-dominated solutions or solutions do not converge to the true Pareto front, then the DW value is very large. In the case of NSGAII and MOPSO for ZDT6, because some points are far away from the true optimal pare fronts, their DW values are 10^{-2}, but the values are 10^{-4} for MOIPSO and we can find that all the solutions are near the true optimal Pareto fronts in Figure 4. Thus, the DW indicator can measure the distribution width of the non-dominated solution set.

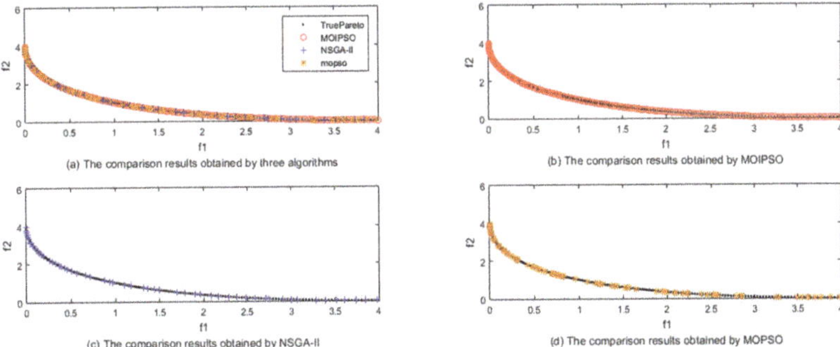

Figure 3. For SCH, the comparisons between the true Pareto front and the best ones obtained by three different algorithms.

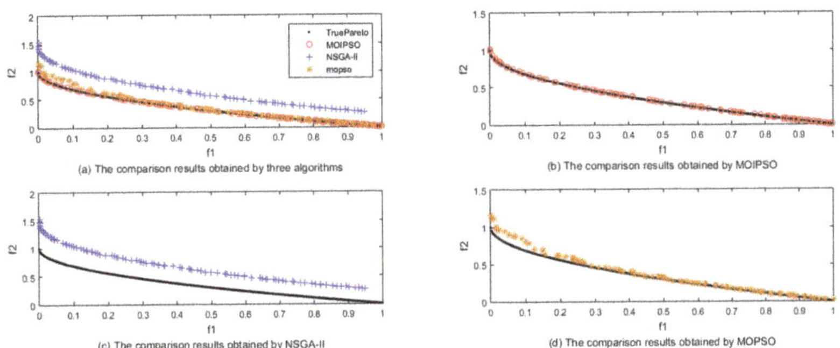

Figure 4. For ZDT1, the comparisons between the true Pareto front and the best ones obtained by three different algorithms.

From the numerical results, we can also see that the MOIPSO algorithm performs better than other algorithms in distribution uniformity and width. This is obviously a good result of the Gaussian mutation throw point strategy. However, this method is not omnipotent. Firstly, when the number of throwing points is too large, the running time will rapidly increase. It is difficult to determine the number, so in this paper, we have done a lot of experiments to determine it. Secondly, for all sparse solutions, we will throw points, so when the Pareto optimal front of optimization problem is very complex and contains a large number of outliers, the effectiveness of our Gaussian mutation throw point strategy will be affected.

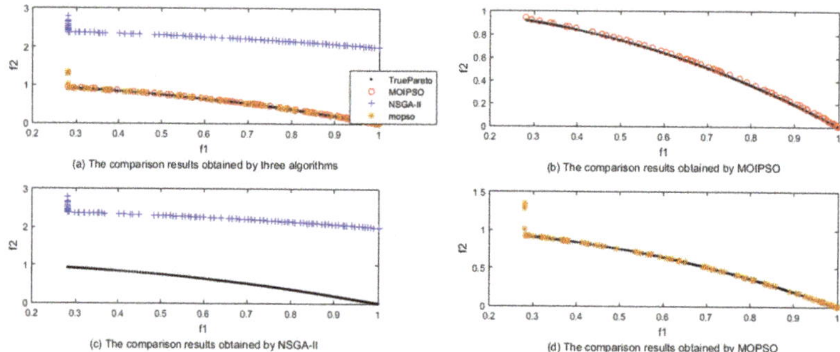

Figure 5. For ZDT6, the comparisons between the true Pareto front and the best ones obtained by three different algorithms.

6. Conclusions

A new multi-objective PSO algorithm based on Gaussian mutation and an improved learning strategy (MOIPSO) is presented to solve MOPs. First, MOIPSO builds different learning strategies to update the individual positions of the non-dominated and dominated solutions. Then, the Gaussian mutation strategy is used to create throw points at sparse and boundary positions. These updating strategies yield high convergence and a satisfactory distribution. To further measure the distribution width, the indicator *DW* is proposed.

The performance of MOIPSO was tested based on different MOP benchmark functions with convex and nonconvex objection functions. To demonstrate the effectiveness of MOIPSO, the results were compared to those of MOPSO and NSGA-II. The experimental results showed that MOIPSO significantly outperforms all other algorithms based on the test problems with respect to three metrics. The resulting data and figures indicate that the proposed *DW* indicator is reasonable.

In this article, only two-objective functions are tested. In the near future, we also plan to evaluate MOIPSO using other objective test functions. Furthermore, most parameters in this paper (such as the number of throw points, cognitive coefficient, etc.) have certain values. It would be interesting to study whether these control parameters could adaptively change as the iteration time increases.

Author Contributions: Both authors have contributed equally to this paper. Both authors have read and approved the final manuscript.

Funding: This research was funded by the National Natural Science Foundation of P.R. China (61561001) and First-Class Disciplines Foundation of NingXia (Grant No. NXYLXK2017B09).

Conflicts of Interest: The authors declare no conflict of interest.

Abbreviations

The following abbreviations are used in this manuscript:

MOIPSO	Multi-objective improved PSO algorithm
MOPs	Multi-objective optimization problems
MOEA	Multi-objective optimization evolutionary algorithm
GMPS	Gaussian mutation points set
TPS	Thickened point set
DW	Distribution width

References

1. Miettinen, K.M. *Nonlinear Multi-Objective Optimization*; Kluwer Academic Publishers: Boston, MA, USA, 1999.
2. Deb, K. *Multi-Objective Optimization Using Evolutionary Algorithms*; John Wiley and Sons: Hoboken, NJ, USA, 2001.
3. Fonseca, C.M.; Fleming, P.J. *Genetic Algorithm for Multi-Objective Optimization: Formulation, Discussion and Generalization*; Morgan Kaufmann Publishers: Burlington, MA, USA, 1993.
4. Deb, K.; Pratap, A.; Agrawal, S.; Meyarivan, T. A fast and elitist multi-objective genetic algorithm: NSGA-II. *IEEE Trans. Evol. Comput.* **2002**, *6*, 182–197. [CrossRef]
5. Zitzler, E.; Thiele, L. Multi-objective evolutionary algorithms: A comparative case study and the strength pareto approach. *IEEE Trans. Evol. Comput.* **1999**, *3*, 257–271. [CrossRef]
6. Knowles, J.D.; Corne, D.W. Approximating the nondominated front using the Pareto archived evolution strategy. *IEEE Trans. Evol. Comput.* **2000**, *8*, 149–172. [CrossRef]
7. Coello, C.A.C.; Pultdo, G.T.; Lechuga, M.S. Handling multiple objectives with particle swarm optimization. *IEEE Trans. Evol. Comput.* **2004**, *8*, 256–279. [CrossRef]
8. Sierra, M.R.; Coello, C.A.C. *Improving PSO-Based Multi-Objective Optimization Using Crowding, Mutation and ε-Dominance*; Springer: Berlin/Heidelberg, Germany, 2005; pp. 505–519.
9. Reddy, M.J.; Kumar, D.N. An efficient multi-objective optimization algorithm based on swarm intelligence for engineering design. *Eng. Optim.* **2007**, *39*, 49–68. [CrossRef]
10. Leong, W.F.; Yen, G.G. PSO-based multi-objective optimization with dynamic population size and adaptive local archives. *IEEE Trans. Syst. Man Cybern. Part B* **2008**, *38*, 1270–1293. [CrossRef] [PubMed]
11. Yen, G.G.; Leong, W.F. Dynamic multiple swarms in multi-objective particle swarm optimization. *IEEE Trans. Syst. Man Cybern. Part B* **2009**, 39, 890–911. [CrossRef]
12. Chen, D.B.; Zou, F.; Wang, J.T. A multi-objective endocrine PSO algorithm and application. *Appl. Soft Comput.* **2011**, *11*, 4508–4520. [CrossRef]
13. Lin, Q.; Li, J.; Du, Z.; Chen, J.; Ming, Z. A novel multi-objective particle swarm optimization with multiple search strategies. *Eur. J. Oper. Res.* **2015**, *247*, 732–744. [CrossRef]
14. Meza, J.; Espitia, H.; Montenegro, C.; Giménez, E.; González-Crespo, R. Movpso: Vortex multi-objective particle swarm optimization. *Appl. Soft Comput.* **2017**, *52*, 1042–1057. [CrossRef]
15. Hu, W.; Yen, G.G. Adaptive multi-objective particle swarm optimization based on parallel cell coordinate system. *IEEE Trans. Evol. Comput.* **2015**, *19*, 1–18.
16. Knight, J.T.; Singer, D.J.; Collette, M.D. Testing of a spreading mechanism to promote diversity in multi-objective particle swarm optimization. *IEEE Trans. Evol. Comput.* **2015**, *16*, 279–302. [CrossRef]
17. Cheng, T.; Chen, M.; Fleming, P.J.; Yang, Z.; Gan, S. A novel hybrid teaching learning based multi-objective particle swarm optimization. *Neurocomputing* **2017**, *222*, 11–25. [CrossRef]
18. Coello, C.A.C. Evolutionary multi-objective optimization: A historical view of the field. *IEEE Comput. Intell. Mag.* **2006**, *1*, 28–36. [CrossRef]
19. Zitzler, E. *Evolutionary Algorithm for Multiobjective Optimization: Methods and Application*; Swiss Federal Institute of Technology: Zurich, Switzerland, 1999.
20. Kennedy, J.; Eberhart, R.C. Particle Swarm Optimization. In Proceedings of the IEEE International Conference on Neural Networks, Perth, Australia, 27 November–1 December 1995; pp. 1942–1948.
21. Wang, Y.N.; Wu, L.H.; Yuan, X.F. Multi-objective self-adaptive differential evolution with elitist archive and crowding entropy-based diversity measure. *Soft Comput.* **2010**, *14*, 193–209. [CrossRef]
22. Higashi, N.; Iba, H. *Particle Swarm Optimization with Gaussian Mutation*; IEEE: Piscataway, NJ, USA, 2013.
23. Coelho, L.D.S.; Ayala, V.H.; Alotto, P. A Multiobjective Gaussian Particle Swarm Approach Applied to Electromagnetic Optimization. *IEEE Trans. Mag.* **2010**, *46*, 3289–3292. [CrossRef]
24. Liang, H.M.; Zhang, K.; Yu, H.H. Multi-objective Gaussian particle swarm algorithm optimization based on niche sorting for actuator design. *Adv. Mech. Eng.* **2015**, *7*, 1–7. [CrossRef]
25. Schaffer, J.D. Multiple objective optimization with vector evaluated genetic algorithm. In Proceedings of the 1st International Conference on Genetic Algorithm and Their Applications, Pittsburg, CA, USA, 24–26 July 1985; pp. 93–100.
26. Kursawe, F. A varant of evolution strategies for vector optimization. In *International Conference on Parallel Problem Solving from Nature*; Springer: Berlin/Heidelberg, Germany, 1991; pp. 193–197.

27. Fonseca, C.M.; Fleming, P.J. Multiobjective optimization and multiple constraint handling with evolutionary algorithms, Part II: Application example. *IEEE Trans. Syst. Man Cybern. A* **1998**, *28*, 38–47. [CrossRef]
28. Zitzler, E.; Deb, K.; Thiele, L. Comparison of multi-objective evolutionary algorithms: Empirical results. *Evol. Comput.* **2000**, *8*, 173–195. [CrossRef]
29. Van Veldhuizen, D.A.; Lamont, G.B. Evolutionary computation and convergence to a Pareto front. In *Late Breaking Papers at the Genetic Programming 1998 Conference*; Standford University: Stanford, CA, USA, 1998; pp. 221–228.
30. Saraswat, A.; Saini, A. Multi-objective optimal reactive power dispatch considering voltage stability in power systems using HFMOEA. *Eng. Appl. Artif. Intell.* **2013**, *26*, 390–404. [CrossRef]
31. Ding, S.X.; Chen, C.; Xin, B.; Psrdalos, P. A bi-objective load balancing model in a distributed simulation system using NSGA-II and MOPSO approaches. *Appl. Soft Comput.* **2018**, *63*, 249–267. [CrossRef]

Article

Dynamic Horizontal Union Algorithm for Multiple Interval Concept Lattices

Yafeng Yang [1], Ru Zhang [2] and Baoxiang Liu [1,*

[1] College of Science, North China University of Science and Technology, 21 Bohai Road, Tangshan 063210, China; hblgyyf@foxmail.com
[2] Department of mathematics and information sciences, Tangshan Normal University, No. 156 Jianshe North Road, Tangshan 063009, China; zhangru1266@126.com
* Correspondence: www1673@163.com

Received: 3 January 2019; Accepted: 29 January 2019; Published: 10 February 2019

Abstract: In the era of big data, the data is updating in real-time. How to prepare the data accurately and efficiently is the key to mining association rules. In view of the above questions, this paper proposes a dynamic horizontal union algorithm of multiple interval concept lattices under the same background of the different attribute set and object set. First, in order to ensure the integrity of the lattice structure, the interval concept lattice incremental generation algorithm was improved, and then interval concept was divided into existing concept, redundancy concept and empty concept. Secondly, combining the characteristics of the interval concept lattice, the concept of consistency of interval concept lattice was defined and it is necessary and sufficient for the horizontal union of the lattice structure. Further, the interval concepts united were discussed, and the principle of horizontal unions was given. Finally, the sequence was scanned by the traversal method. This method increased the efficiency of horizontal union. A case study shows the feasibility and efficiency of the proposed algorithm.

Keywords: big data; interval concept lattice; horizontal union; sequence traversal

1. Introduction

With the era of big data, the complexity of data processing in time and space have become increasingly demanding. Real-time updating of data requires efficient processing of dynamic data. The concept lattice is a powerful tool for data analysis which was proposed by Professor Wille R in 1982 [1]. It has completeness and accuracy, and has been widely applied in information retrieval, digital library, knowledge discovery, and so on [2–4]. Domestic and overseas scholars have carried out various research on concept lattices that mainly include a construction algorithm and improvement [5–7], rule mining based on concept lattices [8,9], and the fusion of other theories such as fuzzy theory, predicate logic, and rough set theory.

For different needs, some expanded concept lattices have been produced, such as, the fuzzy concept lattice, weighted concept lattice, constraint concept lattice, quantitative concept lattice, expansion concept lattice, rough concept lattice etc. [10–12]. In particular, the classic concept lattice, fuzzy concept lattice and weighted concept lattices, the extent contains the objects which meet all the attributes in the intent. To find the concepts which have partial attributes, we must scan the concept lattice and combine the concepts. The time cost is large especially for large concept lattices. While, in the rough concept lattice, although the concepts which have partial attributes can be searched, there may be a lot of objects which only have an attribute of the intent, thus the support and degree of confidence of constructing association rules will be greatly reduced. In practical applications, we often care about the object set which has a certain number or percent of attributes in intent. Then, some pertinent association rules will be mined through a correlation analysis.

Based on the above questions, the interval concept lattice [13] was put forward in 2012 as a collection of objects which had a certain number or percentage of attributes in the connotation. Its expression of concept is (M^α, M^β, Y). Classical concept lattices are constructed from all attributes with full connotations of extensions, and its expression of concept is (X, Y). Rough concept lattices are composed of the concepts of a maximum attribute set and minimum attribute set containing connotative attributes respectively by upper approximate extension and lower approximate extension. The conceptual form of rough concept lattices is (M, N, Y). Interval concept lattices degenerate into classical concept lattices or rough concept lattices when the parameters are $\alpha = 0, \beta = 1$ or $\alpha = 1/|Y|, \beta = 1$, separately. As will be readily seen, interval concept lattice is a general form of classical concept lattice and rough concept lattice.

On the other hand, with the dynamic change of data, the structure of an interval concept lattice will also change. The association rules are also updated in real time. For example, every day a huge amount of transaction information is generated. If we build the interval concept lattice from the daily trading information, we can only tap the local association rules [14,15]. It cannot provide timely and accurate decision-making for the decision-makers from the overall supermarket shopping system. Therefore, it is very necessary to carry out research on the uniting of interval concept lattices to realize data aggregation. Therefore, the dynamic uniting of interval concept lattices is of great significance.

At present, the main union algorithms of concept lattices are as follows: Reference [16] arranges sub-concept lattices vertically or horizontally in ascending or descending order according to the connotation or extension of the concept; Reference [17] proposes synonymous concepts and updates all father-child nodes according to the relationship between father-child concepts; Reference [18] gives an ordered outline by discussing the relationship between the same concept lattices with the same object set. Since the interval concept lattice has only been proposed for about two years, research on it is limited to the progressive generation algorithm of lattice structure, dynamic compression algorithm and association rules mining algorithm. There is no relevant literature on the uniting of interval concept lattices. The uniting of interval concept lattices is divided into vertical and horizontal uniting. The principle and algorithm of vertical uniting of lattices [19] are studied preliminarily. In this paper, the algorithm for dynamic horizontal uniting of interval concept lattices generated from multiple databases was studied.

The structure is as follows: Section 2 introduces the basic concepts of interval concept lattices, the incremental generation algorithm of interval concept lattices [20] and the related concepts of formal context uniting [21,22]; Section 3 proposes the basic theorem of interval concept lattices' horizontal uniting. On the basis of improving the incremental generation algorithm of interval concept lattices, the horizontal uniting algorithm of interval concept lattices was designed. In Section 4, an example is given to demonstrate the feasibility and efficiency of the algorithm.

2. Theoretical and Methodological Basis

For example, in the supermarket shopping system, the promotional manager often pays more attention to the customers who purchase k (k > 1) kinds of goods or more and the potential demand of these customers, and then carry out product marketing to get the greatest benefit through minimum promotion. However, in the existing concept lattice structure, this kind of query cannot be operated directly, and some union connections or filtrations must be performed. The time and space costs are too high. In order to address this problem, interval concept lattices are required.

2.1. Interval Concept Lattice

Definition 1. *For the formal context* (U, A, R) *and its rough concept lattice* $RL(U, A, R)$, (M, N, Y) *is the rough concept. Set an interval* $[\alpha, \beta]$ $(0 \leq \alpha \leq \beta \leq 1)$, *then* α *upper bound extension* M^α *and* β *lower bound extent* M^β *are:*

$$M^\alpha = \{x | x \in M, |f(x) \cap Y|/|Y| \geq \alpha, 0 \leq \alpha \leq 1\} \tag{1}$$

$$M^\beta = \{x | x \in M, |f(x) \cap Y| / |Y| \geq \beta, 0 \leq \alpha \leq \beta \leq 1\} \tag{2}$$

Y is the concept intension and $|Y|$ is the number of elements in Y, that is base number, M^α refers to the objects which may be covered by $\alpha \times |Y|$ attributes or more in Y. M^β refers to the objects which may be covered by $\beta \times |Y|$ attributes or more in Y.

Definition 2. *Suppose (U, A, R) is a formal context and (M^α, M^β, Y) is an interval concept. Y is the intent. M^α is the α upper bound extension and M^β is the β lower bound extension.*

Definition 3. *Suppose (U, A, R) has two interval concepts, $(M_1^\alpha, M_1^\beta, Y_1)$ and $(M_2^\alpha, M_2^\beta, Y_2)$. If $(M_1^\alpha, M_1^\beta, Y_1)$ and $(M_2^\alpha, M_2^\beta, Y_2)$ meet $Y_1 \subseteq Y_2$, $|Y_2| - |Y_1| = 1$, $M_1^\alpha = M_2^\alpha$ and $M_1^\beta = M_2^\beta$, then $(M_1^\alpha, M_1^\beta, Y_1)$ is called the redundant concept.*

Definition 4. *Suppose (U, A, R) has an interval concept, (M^α, M^β, Y). If it meets $M^\alpha = M^\beta = \varnothing$, then (M^α, M^β, Y) is called the empty concept.*

Definition 5. *Suppose (U, A, R) has an interval concept, $C = (M^\alpha, M^\beta, Y)$. If C is neither the redundant concept nor the empty concept, then C is called the existence concept. $L_\alpha^\beta(U, A, R)$ is a collection of all the existence concepts.*

Definition 6. *$\overline{L_\alpha^\beta(U, A, R)}$ refers to all the $[\alpha, \beta]$ interval concepts, which includes: existence concepts, redundant concepts and empty concepts, that is:*

$$(M_1{}^\alpha, M_1{}^\beta, Y_1) \leq (M_2{}^\alpha, M_2{}^\beta, Y_2) \Leftrightarrow Y_1 \supseteq Y_2, \tag{3}$$

Then "$\leq$" is called the partial order relationship of $\overline{L_\alpha^\beta(U, A, R)}$.

Definition 7. *If all the concepts in $\overline{L_\alpha^\beta(U, A, R)}$ meet "$\leq$", then $L_\alpha^\beta(U, A, R)$ calls interval concept lattice on the formal context (U, A, R).*

Definition 8. *In the interval concept lattice $L_\alpha^\beta(U, A, R)$, if $C = (M^\alpha, M^\beta, Y) \in L_\alpha^\beta(U, A, R)$, then the layers of the Lattice Structure is $|A| + 1$ and node C is at Layer $|Y|$. In particular, when $Y = \varnothing$, C was recorded on the zeroth layer.*

2.2. Incremental Construction Algorithm of Interval Concept Lattice (ICAICL)

Thought of algorithm [11] is as follows.

(1) Calculate the attribute power set $P(A)$ from formal context and let each element Y of power set be intent. Construct the initial node-set G according to the intension cardinal number in ascending order. For clarity, suppose each concept is a six-point group:

$$(M^\alpha, M^\beta, Y, Parent, Children, No), \tag{4}$$

where M^α, M^β is null set, and $Parent = Children = "NULL"$.

(2) Set parameters α, β and the node of Y is $G = (M^\alpha, M^\beta, Y)$. Traversing the intent Y_i of every object, if $Y_i \subseteq Y$ and $|Y_i|/|Y| \geq \alpha$, then merge the object into M^α; if $Y_i \subseteq Y$ and $|Y_i|/|Y| \geq \beta$, merge the object into M^β.

(3) Firstly, construct the root and end node, then insert other nodes as new ones into the lattice incrementally to form the lattice structure. After inserting a new one, there are three kinds of nodes in the $L_\alpha^{\beta\prime}$: New node (the node inserted), invariant and update node.

3. Dynamic Horizontal Union Algorithm

3.1. Basic Principles

Definition 9. *In the two formal context $K_1 = (G_1, M_1, I_1)$ and $K_2 = (G_2, M_2, I_2)$, if $gI_1m \Leftrightarrow gI_2m$ for each $g \in G_1 \cap G_2$ and $m \in M_1 \cap M_2$, then K_1 and K_2 are consistent [12,13]. Otherwise, K_1 and K_2 are inconsistent.*

Definition 10. *Suppose the formal context (U_1, A_1, R_1) and (U_2, A_2, R_2) is consistent and they compose interval concept lattice $L_{\alpha_1}^{\beta_1}(U_1, A_1, R_1)$ and $L_{\alpha_2}^{\beta_2}(U_2, A_2, R_2)$ respectively. When they meet $\alpha_1 = \alpha_2 = \alpha$ and $\beta_1 = \beta_2 = \beta$, then $L_\alpha^\beta(U_1, A_1, R_1)$ and $L_\alpha^\beta(U_2, A_2, R_2)$ call consistently, otherwise they are inconsistent.*

Definition 11. *If interval concept lattice $L_\alpha^\beta(U_1, A_1, R_1)$ and $L_\alpha^\beta(U_2, A_2, R_2)$ is consistent and meet $U = U_1 = U_2$, we get $L_\alpha^\beta(U, A, R)$ when we combine them. $L_\alpha^\beta(U, A, R)$ call horizontal union of $L_\alpha^\beta(U_1, A_1, R_1)$ and $L_\alpha^\beta(U_2, A_2, R_2)$. Suppose $C_1 \in L_\alpha^\beta(U_1, A, R_1)$, $C_2 \in L_\alpha^\beta(U_2, A, R_2)$, then $C = C_1 \otimes C_2$ is the horizontal union of the interval concept.*

The concepts in $L_\alpha^\beta(U, A, R)$ can be divided into two categories. The first is that the number of conceptual connotative attributes in the original structure equals the number of layers of the concepts, and the second is that the sum of the number of conceptual connotative attributes in the two original structures equals the number of layers of the concepts. The specific combination of the two concepts is as follows:

Theorem 1. *Suppose $L_\alpha^\beta(U_1, A_1, R_1)$ and $L_\alpha^\beta(U_2, A_2, R_2)$ is consistent and meet $A_1 = A_2 = A$. At the same time suppose $(M_1^\alpha, M_1^\beta, Y_1)$ and $(M_2^\alpha, M_2^\beta, Y_2)$ are interval concepts in $L_\alpha^\beta(U_1, A, R_1)$ and $L_\alpha^\beta(U_2, A, R_2)$ respectively. If they meet $Y = Y_1 = Y_2$, $M^\alpha = M_1^\alpha \cup M_2^\alpha$ and $M^\beta = M_1^\beta \cup M_2^\beta$, then (M^α, M^β, Y) is the interval concept after vertical union.*

Theorem 1 shows that for the first kind of concepts, the extension and intension of upper and lower bounds of the concepts themselves remain unchanged before and after uniting. But the flags of conceptual types are different. When the conceptual labels flag equals 2 and flag equals 3 in the lattice structure, they remain unchanged after uniting; when the conceptual label flag equals 1 in the lattice structure, flags may equal 1 or 2 after uniting.

Proof. Since the formal context corresponding to the primitive structure is contained in the formal context corresponding to the united lattice structure, the ternary ordered pairs of concepts in the primitive structure are completely preserved in the united lattice structure. At the same time, there is no change before and after the uniting of empty concepts and redundant concepts. For the existence concepts, the new concepts generated by uniting may make themselves redundant.

For the second kind of concept, $C = (M^\alpha, M^\beta, Y) \in L_\alpha^\beta(U, A, R)$, $C_1 = (M_1^\alpha, M_1^\beta, Y_1) \in L_\alpha^\beta(U, A_1, R_1)$, $C_2 = (M_2^\alpha, M_2^\beta, Y_2) \in L_\alpha^\beta(U, A_2, R_2)$. When $Y_1 \cap Y_2 = \varnothing$, taking the upper boundary extension as an example, the case of horizontal union can be divided into the following two types: See Theorems 2 and 3. $\square$

Theorem 2. *When $\lceil |Y_1|\alpha \rceil + \lceil |Y_2|\alpha \rceil = \lceil |Y_1 \cup Y_2|\alpha \rceil$, $M^\alpha = M_1^\alpha \cap M_2^\alpha$, verify whether $x_1 = M_1^\alpha - M^\alpha$ and $x_2 = M_2^\alpha - M^\alpha$ are empty, if not empty, then bring them into and verify whether $\frac{|f(x) \cap (Y_1 \cup Y_2)|}{|Y|} \geq \alpha$. If $\frac{|f(x) \cap (Y_1 \cup Y_2)|}{|Y|} \geq \alpha$, add objects to M^α, or not. Symbol $\lceil x \rceil$ denotes the smallest integer greater than or equal to x.*

Proof. Determined by the boundary extension itself $\frac{|f(x) \cap Y|}{|Y|} \geq \alpha$, the object satisfying this formula contains at least B attributes, and $Y_1 \cap Y_2 = \varnothing$. $\lceil |Y_1|\alpha \rceil + \lceil |Y_2|\alpha \rceil = \lceil |Y_1 \cup Y_2|\alpha \rceil$ is equivalent to

$M^\alpha = M_1^\alpha \cap M_2^\alpha$. However, $x_1 = M_1^\alpha - M^\alpha$ or $x_2 = M_2^\alpha - M^\alpha$ may still contain objects larger than or equal to $(|Y_1 \cup Y_2|\alpha)$, so it needs further verification and cannot be eliminated directly. $\square$

Theorem 3. *When* $\lceil |Y_1|\alpha \rceil = \lceil |Y_1 \cup Y_2|\alpha \rceil$, $M^\alpha = M_1^\alpha$, *verify whether* $x_1 = M_1^\alpha - M^\alpha$ *and* $x_2 = M_2^\alpha - M^\alpha$ *are empty, if not empty, then bring them into and verify whether* $\frac{|f(x) \cap (Y_1 \cup Y_2)|}{|Y|} \geq \alpha$. *If* $\frac{|f(x) \cap (Y_1 \cup Y_2)|}{|Y|} \geq \alpha$, *adds objects to* M^α, *or not.* $\lceil |Y_2|\alpha \rceil = \lceil |Y_1 \cup Y_2|\alpha \rceil$.

Theorem 4. *Supposes that* $L_\alpha^\beta(U, A, R)$ *is the interval concept lattice obtained by the horizontal uniting of* $L_\alpha^\beta(U, A_1, R_1)$ *and* $L_\alpha^\beta(U, A_2, R_2)$ $(A_1 \cap A_2 = \varnothing)$, *then the m-level node of* $L_\alpha^\beta(U, A, R)$ *is* $C_{|A_1|+|A_2|}^m =$ $C_{|A_1|}^m + C_{|A_1|}^{m-1}C_{|A_2|}^1 + C_{|A_1|}^{m-2}C_{|A_2|}^2 + \ldots + C_{|A_1|}^1 C_{|A_2|}^{m-1} + C_{|A_2|}^m$.

Here $C_{|A_1|}^m$ refers to all interval concepts whose number of intension attributes is m in $L_\alpha^\beta(U, A_1, R_1)$, $C_{|A_1|}^{m-1}C_{|A_2|}^1$ refers to all interval concepts whose number of intension attributes is $m - 1$ in $L_\alpha^\beta(U, A_1, R_1)$ and all interval concepts whose number of intension attributes is 1 in $L_\alpha^\beta(U, A_2, R_2)$ merge horizontally.

3.2. Algorithmic Design

In order to generate interval concept lattices while retaining all interval concepts, including existing concepts, redundant concepts and empty concepts, and effectively improve the uniting efficiency of interval concept lattices, the existing incremental interval concept lattice generation algorithm needs to be modified first, and on this basis, a dynamic horizontal uniting algorithm of multiple interval concept lattices is proposed.

3.2.1. Improved Progressive Generation Algorithms for Interval Concept Lattices

In order to distinguish between different interval concepts, concept nodes are defined and stored in structured form as follows:

$$flag \;\middle|\; M^\alpha \;\middle|\; M^\beta \;\middle|\; Y \;\middle|\; parent \;\middle|\; children \;\middle|\; no$$

It is defined in the following form.

struct concept

{ string M_i^α; $M_i^\alpha M_i^\beta$; Y_i;

struct Y, parent, children;

int flag; }

The concept of category is marked by flag. When flag = 1, flag = 2 and flag = 3, stored concept is exist concept, redundant concept and empty concept separately.

Algorithm 1. Improved ICAICL

Input: formal context (U, A, R)

Output: interval concept lattice L_α^β and $\overline{L_\alpha^\beta}$

(1) Calculate power set of attribute set P(A) to determine the intent of concept, generate concept node of initialization G.

(2) Determine upper bound extent M_i^α and lower bound extent M_i^β, make flag of empty concept be 3 and the other be 1.

(3) According to the partial order relationship, determine the level of the node and the parent-child relationship, make flag of redundant concept be 2.

The method of finding out redundant concept is Romove-redun (Ch,Gi).

Remove-redun(Ch,Gi) //find out redundant concept, sign and store, and delete it

{   for each children Ch in Gi // Ch pointer point to every children of Gi

{ If (Gi.M_i^α = Ch.M_i^α, Gi.M_i^β = Ch.M_i^β)

{   Flag=2;

Delete Gi from L_α^β}}}

(4) To find concept of no = 1, structure the root node. Then insert other nodes into the lattice according to the parent-child relationship successively. Eventually form the structure of interval concept lattice.

3.2.2. Dynamic Lateral Uniting Algorithms for Multi-Interval Concept Lattices

The basic idea of the algorithm is that whenever a new data set is generated, it will be transformed into a formal representation of the interval concept lattice and united with the united lattice structure. In this way, the information of the interval concept lattice can be aggregated again and again, which lays a foundation for further mining association rules.

Algorithm 2. DHM (Dynamic Horizontal union)

Input: $L_\alpha^\beta 1, \overline{L_\alpha^\beta} 1, L_\alpha^\beta 2, \overline{L_\alpha^\beta} 2 \ldots L_\alpha^\beta i, \overline{L_\alpha^\beta} i \ldots$

Output: L_α^β and $\overline{L_\alpha^\beta}$

Steps:

(1) Let $L_\alpha^\beta = L_\alpha^\beta 1$ and $\overline{L_\alpha^\beta} = \overline{L_\alpha^\beta} 1$;

(2) The newly generated interval concept lattices $L_\alpha^\beta i$ and $\overline{L_\alpha^\beta} i$ $(i = 2, 3 \ldots n)$ are united with L_α^β and $\overline{L_\alpha^\beta}$ respectively. The united results are assigned to L_α^β and $\overline{L_\alpha^\beta}$. The uniting steps of the two interval concept lattices are as follows:

① Let $A_i^\wedge = A_i \cap A$, delete the concept of attributes in the newly generated interval concept lattice $L_\alpha^\beta(U, A_i, R_i)(L_\alpha^\beta i)$ containing duplicate concept attributes set $A_i^\wedge$, and mark its lattice structure as $L_\alpha^\beta(U, A_i^*, R_i^*)(L_\alpha^\beta i^*)$ and corresponding attributes set as $A_i^* = A_i - A_i^\wedge$.

② Let $A^* = A \cup A_i^*$, compute the intension of the concept determined by attribute set $P(A^*)$, and generate concept node set G^* of the initialized concept $\overline{L_\alpha^{\beta^*}}$. According to node set G^*, the hierarchy and parent-child relationship of the node are determined according to partial order relation. Let flag = 0, and the upper and lower bounds are empty.

③ Scanning the interval concepts in $\overline{L_\alpha^\beta}$ and $\overline{L_\alpha^\beta} i^*$ by sequence and generating the concepts of $\overline{L_\alpha^{\beta^*}}$. There, Horizontal union-M Layer is used for horizontal uniting of layer M.

Horizontal union-M Layer $(\overline{L_\alpha^\beta} i^*, C_i^*, \overline{L_\alpha^\beta}, C, \overline{L_\alpha^{\beta^*}}, C^*)$

{   For C_i^* in $\overline{L_\alpha^\beta} i^*$;

For C in $\overline{L_\alpha^\beta}$;

If $(C_i^*.|Y| = m)$ // the intention attributes number of the concept in $\overline{L_\alpha^\beta} i^*$ is equal to the concept layer m

{   If $(C^*.Y = C_i^*.Y)$

{ $C^*.flag = C_i^*.flag$;

$C^*.M^\alpha = C_i^*.M^\alpha$;

$$C_1^*.M^\beta = C_i^*.M^\beta; \}$$

Else end }

If $(C.|Y| = m)$ // the intention attributes number of the concept in $\overline{L_\alpha^\beta}$ is equal to the concept layer m
    {   If $(C^*.Y = C.Y)$
{ $C^*.flag = C.flag;$
         $C^*.M^\alpha = C.M^\alpha;$
         $C^*.M^\beta = C.M^\beta; \}$

Else end }

If $(C_i^*.|Y| + C.|Y| = m)$ // The sum of the number of attributes in $\overline{L_\alpha^\beta i^*}$ and $\overline{L_\alpha^\beta}$ is equal to the concept
layer m
    { For $(C_i^*.|Y| = 1; C_i^*.|Y| = m - 1; C_i^*.|Y| + +)$
        $C.|Y| = m - C_i^*.|Y|$
For $C^*.M^\alpha$
If $(\lceil C_i^*.|Y| * \alpha \rceil + \lceil C.|Y| * \alpha \rceil = \lceil (C_i^*.|Y| + C.|Y|) * \alpha \rceil)$
{ $\gamma = \alpha$
            union1-$C_i^*C(C_i^*, C, C^*)$ }
              If $(\lceil C_i^*.|Y| * \alpha \rceil = \lceil (C_i^*.|Y| + C.|Y|) * \alpha \rceil$ or $\lceil C.|Y| * \alpha \rceil = \lceil (C_i^*.|Y| + C.|Y|) * \alpha \rceil)$
{   If $\lceil C_i^*.|Y| * \alpha \rceil = \lceil (C_i^*.|Y| + C.|Y|) * \alpha \rceil$
{ $\gamma = \alpha;$
        $C_g = C_i^*;$
        $C_h = C;$
            union2-$C_g C_h(C_g, C_h, C^*)$}
If $(\lceil C.|Y| * \alpha \rceil = \lceil (C_i^*.|Y| + C.|Y|) * \alpha \rceil$
{ $\gamma = \alpha;$
        $C_g = C;$
        $C_h = C_i^*;$
            union2-$C_g C_h(C_g, C_h, C^*)$}
          Else end }}
For $C^*.M^\beta$
   If $(\lceil C_i^*.|Y| * \beta \rceil + \lceil C.|Y| * \beta \rceil = \lceil (C_i^*.|Y| + C.|Y|) * \beta \rceil)$
{ $\gamma = \beta$
           union1-$C_i^*C(C_i^*, C, C^*)$}
            If $(\lceil C_i^*.|Y| * \beta \rceil = \lceil (C_i^*.|Y| + C.|Y|) * \beta \rceil$ or $\lceil C.|Y| * \beta \rceil = \lceil (C_i^*.|Y| + C.|Y|) * \beta \rceil)$
{   If $(\lceil C_i^*.|Y| * \beta \rceil = \lceil (C_i^*.|Y| + C.|Y|) * \beta \rceil$
{ $\gamma = \beta;$
        $C_g = C_i^*;$
        $C_h = C;$
            union2-$C_g C_h(C_g, C_h, C^*)$}
If $(\lceil C.|Y| * \beta \rceil = \lceil (C_i^*.|Y| + C.|Y|) * \beta \rceil$
{ $\gamma = \beta;$
        $C_g = C;$
        $C_h = C_i^*;$
            union2-$C_g C_h(C_g, C_h, C^*)$}
          Else end }}
Remove-redun(Ch,Gi)
If $(C^*.M^\alpha = C^*.M^\beta = \varnothing)$
Flag = 3
Else flag = 1}}

The Horizontal union-M Layer algorithm calls three sub-functions, Remove-redun(Ch,Gi), union1-$C_i^*C(C_i^*, C, C^*)$ and union2-$C_i^*C(C_i^*, C, C^*)$ respectively. The function of Remove-redun(Ch,Gi) is to find redundant concepts of united interval concept lattices and mark them. The function of union1-$C_i^* C(C_i^*, C, C^*)$ and union2-$C_i^* C(C_i^*, C, C^*)$ are to calculate and test the deleted objects for uniting cases, and to get the concept of interval after uniting.

Assign L_α^β and $\overline{L_\alpha^\beta}$ to $L_\alpha^{\beta*}$ and $\overline{L_\alpha^{\beta*}}$ separately. Let $i = i + 1$, go to step 2.

The algorithm starts from the interval concept lattice directly, makes full use of the lattice structure of the original interval concept lattice and covers the uniting of all concepts. Therefore, it has completeness and effectiveness. Compared with the method of uniting formal context first and then using ICAICL algorithm to construct the interval concept lattice, this algorithm reduces time complexity and has a value of $O(n * n_i * (n + n_i))$, which proves the efficiency of the algorithm.

4. Example Analysis

Two formal contexts are listed in Tables 1 and 2. $U = \{1, 2, 3, 4, 5\}$ is the element set. $A_1 = \{a, b, c\}$ and $A_2 = \{c, d, e, f\}$ are attribute sets before and after the change, respectively. Their corresponding lattice structures and union are described as follows.

(1) Set $\alpha = 0.6$, $\beta = 0.7$, applying the improved interval concept lattice progressive generation algorithm to generate the primitive lattice structure is shown in Figure 1a,b.

(2) $A = \{a, b, c, d, e, f\}$. Flag = 0. The upper and lower boundaries are empty. Initialization of Interval Concept Constitution Generated by Parent-Child Relation is $\overline{L_\alpha^\beta}$.

(3) $A^* = A_1 \cap A_2 \neq \varnothing$, $A^* = c$. Deleting all the concept nodes in $\overline{L_\alpha^\beta}1$ which contain the attribute c. Here, Lattice structure is $\overline{L_\alpha^\beta 1}^*$ and its attributes set is $A_1^* = A_1 - A^*$.

(4) Scanning $\overline{L_\alpha^\beta 1}^*$ and $\overline{L_\alpha^\beta}2$ in sequence, and the interval concept is generated in different cases. L_α^β can be obtained from $\overline{L_\alpha^\beta}$. The interval concept lattice after horizontal union is shown in Figure 2.

Table 1. The formal context of $L_\alpha^\beta 1$.

U \ A₁	a	b	c
1	1	0	1
2	0	1	0
3	1	1	0
4	1	0	0
5	1	0	1

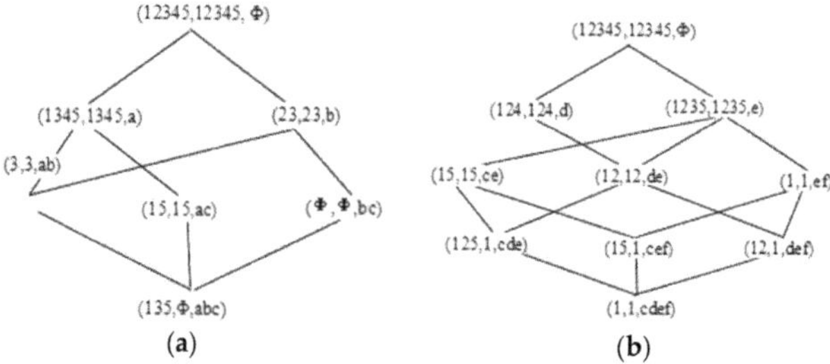

Figure 1. (a) The lattice structure $L_\alpha^\beta(U, A_1, R_1)$ of $L_\alpha^\beta 1$; (b) The lattice structure $L_\alpha^\beta(U, A_2, R_2)$ of $L_\alpha^\beta 2$.

Take Layer 3 as an example, several cases of horizontal uniting of interval concepts are described below. The number of attributes in $\overline{L_\alpha^\beta 1}^*$ and $\overline{L_\alpha^\beta}2$ are 2 and 4. According to $C_6^3 = C_4^3 + C_2^1 C_4^2 + C_2^2 C_4^1$, the uniting of the third layer of L_α^β can be divided into three cases.

- Case 1: There are four concepts of the third layer in $\overline{L_\alpha^\beta 2}$. According to the same principle of intension attributes, the corresponding concepts in $\overline{L_\alpha^\beta}$ are found, and the upper and lower bounds of the concepts and their flags are assigned to the corresponding interval concepts.

- Case 2: In $\overline{L_\alpha^\beta 1}^*$, the nodes which have one intension attribute are $(1345,1345,a)$ and $(23,23,b)$, and the nodes in $\overline{L_\alpha^\beta 2}$ which have two intension attributes are $(1,1,cd)$, $(15,15,ce)$, $(1,1,cf)$, $(12,12,de)$, $(1,1,df)$ and $(1,1,ef)$. Merge them horizontally. Firstly, the uniting case between $(1345,1345,a)$ and $(1,1,cd)$ belongs to $\lceil |cd| * 0.6 \rceil = \lceil (|a| + |cd|) * 0.6 \rceil$, so $M^{\alpha*} = \{1\}$. Meanwhile, $\{1345\} - \{1\} = \{345\} \neq \varnothing$, bring the elements of $\{345\}$ into $|f(x) \cap acd| / |acd| \geq 0.6$, determining whether the proportion of intension attributes corresponding to deleted objects satisfies the relationship further, and then add the object into $M^{\alpha*}$ if the relationship is satisfied, otherwise it will be eliminated completely. By analogy, the upper and lower boundary extensions are used to determine the uniting case, and the deletion concept is further verified. Finally, the type of the deletion concept is determined according to the upper and lower boundary extensions after the concept is generated.

- Case 3: Uniting the nodes which have two intension attributes in $\overline{L_\alpha^\beta 1}^*$ and the nodes which have one intension attributes in $\overline{L_\alpha^\beta 2}$. The interval concepts obtained from the above three cases together constitute the third layer of $\overline{L_\alpha^\beta}$.

Table 2. The formal context of $L_\alpha^\beta 2$.

U \ A₁	c	d	e	f
1	1	1	1	1
2	0	1	1	0
3	0	0	1	0
4	0	1	0	0
5	1	0	1	0

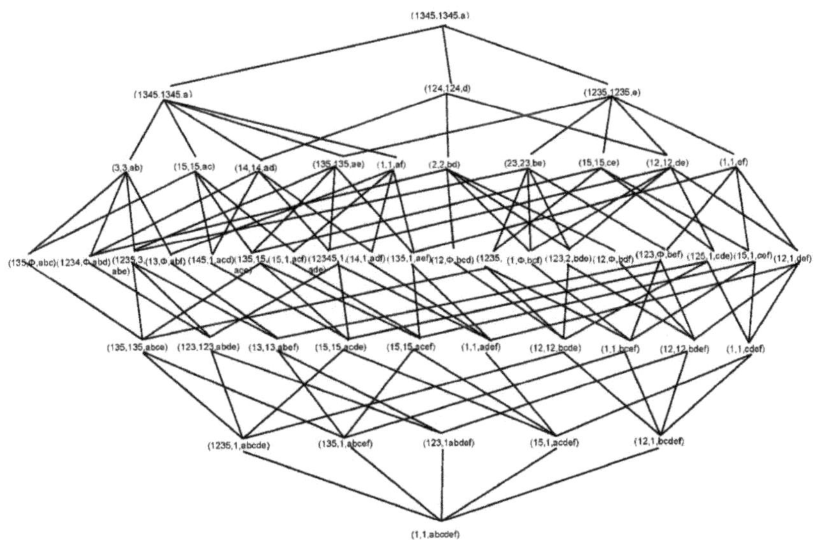

Figure 2. The lattice structure $L_\alpha^\beta(U, A, R)$ united by $L_\alpha^\beta(U, A_1, R_1)$ and $L_\alpha^\beta(U, A_2, R_2)$ for the Interval parameters $\alpha = 0.6$ and $\beta = 0.7$.

Then $\overline{L_\alpha^\beta}$ can be obtained by horizontal uniting between conceptual nodes. After eliminating the interval concepts of flag equaling 2 and 3 in $\overline{L_\alpha^\beta}$, it can be transformed into L_α^β. See Figure 2.

5. Conclusions

When facing real-time data updating in the era of big data, how to deal with data effectively which are generated at any time has become one of the key issues. In this paper, the concept of consistency is introduced as a prerequisite for the uniting of interval concept lattices. Lateral uniting of interval concept lattices can be carried out when the parameter intervals of two interval concept lattices are identical and the object sets are identical. In order to preserve the integrity of the lattice structure, the concepts in lattice structures are divided into three categories: Existential concept, redundant concept and empty concept by improving the progressive generation algorithm of interval concept lattices. The concept of $\overline{L(U, A, R)}$ is introduced at the same time. Lateral uniting of lattice structures is specified to the horizontal uniting of lattice nodes. It can be divided into two situations, that is, the number of connotative attributes of concepts in the original structure equals the number of conceptual layers and the sum of the number of connotative attributes of concepts in the two original structures equals the number of conceptual layers, and the two cases are further refined according to the actual situation, so as to realize the horizontal union of interval concept lattices.

However, in the face of large-scale data, the structure of the interval concept lattice will be greatly expanded, which will lead to high operational complexity. How to develop efficient dynamic uniting software of interval concept lattice, further reduce the complexity of time and space, realize the optimal merging of interval concept lattices, and proposing association rules from them will be the next major research work.

Author Contributions: Y.Y. contributed to the main body of the paper, the improved method of constructing the interval concept lattice, and the principle of dynamic combination. The main models and algorithms were constructed; R.Z. was mainly responsible for the writing of the paper, colleagues performed data processing and graphics drawing; B.L. guided the overall process of the paper, gave the definition of interval concept lattices, and points out the method and feasibility of the combination of interval concept lattices.

Funding: This research was funded by the National Natural Science Foundation of China (61370168, 61472340), Natural Science Foundation of Hebei Province (F2016209344).

Acknowledgments: The authors thank Chunying Zhang, Lihong Li and Bin Bai from North China University of Science and Technology for their technical assistance during materials processing and data acquisition.

Conflicts of Interest: The authors declare no conflicts of interest.

References

1. Wille, R. Restructing lattice theory: An approach based on hierarchies of concepts. In *Ordered Sets*; Rival, I., Ed.; Reidel: Dordrecht, The Netherlands, 1982; pp. 445–470.
2. Godin, R.; Missaoui, R.; April, A. Experimental compare-eison of navigation in a Galois lattice with conventional information retrieval methods. *Int. J. Man-Mach. Stud.* **1993**, *38*, 747–767. [CrossRef]
3. Carpineto, C.; Romano, G. Information retrieval through hybrid navigation of lattice representations. *Int. J. Hum.-Comput. Stud.* **1996**, *45*, 553–578. [CrossRef]
4. Cole, R.; Stumme, G. CEM—A conceptual email manager. In Proceedings of the 7th International Conference on Conceptual Structures (ICCS'2000), Darmstadt, Germany, 14–18 August 2000; Springer Verlag: Berlin/Heidelberg, Germany, 2000.
5. Zhang, C.; Liu, B.; Guo, J.; Liu, F.C. Incremental algorithm for building association rule lattice based on attribute linded-list. *Comput. Eng. Des.* **2005**, *26*, 320–323.
6. Qiang, Y.; Liu, Z.; Lin, W.; Shi, B.; Li, Y. Research on an Algorithm for Fuzzy Concept Lattice Construction. *Comput. Eng. Appl.* **2004**, *29*, 50–53.
7. Fuentes-Gonzlez, B. Construction of the L-fuzzy concept lattice. *Fuzzy Sets Syst.* **1998**, *97*, 109–114.
8. Xie, Z.; Liu, Z. Concept Lattice and Association Rule Discovery. *J. Comput. Res. Dev.* **2000**, *37*, 1415–1421.

9. Qiu, G.; Zhu, Z. Acquisitions to Decision Rules and Algorithms to Inferences Based on Crisp-fuzzy Variable Threshold Concept Lattices. *Comput. Sci.* **2009**, *36*, 216–218.

10. Xie, Z. Knowledge Discovery Based on Concept Lattice Model. Doctoral Dissertation, Hefei University of Technology, Hefei, China, 2001; pp. 1–20.

11. Zhang, J.; Zhang, S. Weighted Concept Lattice and Incremental Construction. *Pattern Recognit. Artif. Intell.* **2005**, *18*, 171–176.

12. Yang, H.; Zhang, J. Rough concept lattice and construction arithmetic. *Comput. Eng. Appl.* **2007**, *43*, 172–175.

13. Liu, B.; Zhang, C. New concept lattice structure-interval concept lattice. *Comput. Sci.* **2012**, *39*, 273–277.

14. Zhai, Y.; Guo, W.; Wang, L. A new association rule classification algorithm using extended concept lattice. *J. Liaoning Tech. Univ. (Nat. Sci.)* **2015**, *34*, 1280–1284.

15. Liu, J.; Liu, B.; Chen, H. Rules extraction of fuzzy association rules lattice. *J. Liaoning Tech. Univ. (Nat. Sci.)* **2013**, *32*, 852–856.

16. Yao, J.; Yang, S.; Li, X.; Peng, Y. Incremental double sequence algorithm of concept lattice union. *Appl. Res. Comput.* **2013**, *30*, 1038–1040.

17. Zhang, L.; Shen, X.; Han, D.; An, G. Vertical union algorithm of concept lattices based on synonymous concept. *Comput. Eng. Appl.* **2007**, *43*, 95–98.

18. Li, H. A vertical union algorithm of concept lattices. *Nat. Sci. J. Harbin Norm. Univ.* **2010**, *26*, 27–31.

19. Zhang, R.; Zhang, C.; Wang, L.; Liu, B. Vertical union algorithm of interval concept lattices. *J. Comput. Appl.* **2015**, *35*, 1217–1221.

20. Zhang, C.; Wang, L. Incremental construction algorithm based on attribute power set for intercal concept lattice L_α^β. *Appl. Res. Comput.* **2014**, *31*, 731–734.

21. Liu, Z.T.; Shen, X.J.; Wu, Q.; Qiang, Y. Theoretical research on the distributed construction of concept lattices. In Proceedings of the Second International Conference on Machine Learning and Cybernetics, Institute of Electrical and Electronics, Xi'an, China, 2–5 November 2003; pp. 474–479.

22. Zhi, H.; Zhi, D.; Liu, Z. Theory and algorithm of concept lattice union. *Acta Electron. Sin.* **2010**, *38*, 455–459.

 *mathematics*

Article

Optimal Task Allocation in Wireless Sensor Networks by Means of Social Network Optimization

Alessandro Niccolai *, Francesco Grimaccia, Marco Mussetta and Riccardo Zich

Dipartimento di Energia, Politecnico di Milano, Via Lambruschini 4, 20156 Milano, Italy;
francesco.grimaccia@polimi.it (F.G.); marco.mussetta@polimi.it (M.M.); riccardo.zich@polimi.it (R.Z.)
* Correspondence: alessandro.niccolai@polimi.it

Received: 5 March 2019; Accepted: 25 March 2019; Published: 28 March 2019

Abstract: Wireless Sensor Networks (WSN) have been widely adopted for years, but their role is growing significantly currently with the increase of the importance of the Internet of Things paradigm. Moreover, since the computational capability of small-sized devices is also increasing, WSN are now capable of performing relevant operations. An optimal scheduling of these in-network processes can affect both the total computational time and the energy requirements. Evolutionary optimization techniques can address this problem successfully due to their capability to manage non-linear problems with many design variables. In this paper, an evolutionary algorithm recently developed, named Social Network Optimization (SNO), has been applied to the problem of task allocation in a WSN. The optimization results on two test cases have been analyzed: in the first one, no energy constraints have been added to the optimization, while in the second one, a minimum number of life cycles is imposed.

Keywords: wireless sensor networks; task allocation; stochastic optimization; social network optimization

---

1. Introduction

Wireless Sensor Networks (WSN) are relevant system architectures that can be applied in a wide range of applications [1], from monitoring to tracking, visual surveillance, ranging also in many fields from industrial automation to agricultural systems, and target localization, both in military and civil sectors [2].

Currently, WSN are becoming much more important because the senors' computational capabilities are growing, and thus, many in-network tasks can be performed. It has been noticed that processing the information inside the network is faster and safer than sending raw data to the final user [3]. Such in-network processing can drastically reduce the total computational time required after sensing tasks with a direct impact on power consumption [4].

The network complex dynamics created by these in-network operations can be suitably managed by means of evolutionary optimization techniques [5]. For instance, the problem of sensor lifetime maximization has been successfully approached in [6] by means of genetic algorithms and in [7] with genetical swarm optimization. The problem of coverage in WSN has been solved in [8] with both the genetic algorithm and ant colony optimization. Another important problem of WSN, routing, has been widely approached by several authors: in [9], a specific energy protocol has been designed to improve the network lifetime, while in [10], it has been solved with genetical swarm optimization and in [11] with particle swarm optimization.

The problem of task and resource allocation is a crucial problem in many frameworks. It is aimed at finding the optimal distribution of the tasks inside the network itself with reference to a specific goal. In the case of the deployment of multiple sensor devices with batteries, a fundamental goal is to maximize the total network lifetime, in other words to minimize the total energy consumption [12]. In many WSN applications, the computational time is also a key parameter in the optimization process. Scheduling and planning have been already faced with evolutionary computation algorithms, like the genetic algorithm [13], particle swarm optimization [14], and ant colony optimization [15]. In standard planning problems, these techniques have been demonstrated to have performances comparable with other deterministic optimization techniques [16].

With recent advancements in massive parallel computing technologies, the problem of scheduling resources and tasks, for example in multiprocessor systems, is becoming more and more attractive. In this scenario, soft computing techniques represent a useful tool that can be effectively applied in task allocation optimization problems.

This work is based on a multi-hop network in which a set of tasks for parallel computing should be performed. The tasks are distributed and collected by a central node. By properly choosing the number of such processed tasks for each sensor, it is possible to change the total processing time and the network lifetime.

The analyzed task allocation problem is devoted to the minimization of the elaboration cycle time of the entire network, i.e., the total time required by the network to process all the tasks. The optimization is performed fixing a predefined scheduling protocol for each sensor and taking into consideration energy constraints expressed in terms of network lifetime.

To perform this optimization, a promising evolutionary optimization algorithm, named Social Network Optimization (SNO) [17], has been used. SNO is a recently-developed population-based algorithm: in the literature, it has been applied to antenna optimization [18], with a comparison between SNO and PSO; it has been also applied to model parameter matching problems [19].

In order to assess SNO performance on task allocation, in this paper, the final results are compared with the solutions obtained by the following algorithms: Biogeography-Based Optimization (BBO), Differential Evolutionary (DE), the Genetic Algorithm (GA), Particle Swarm Optimization (PSO), and the Stud-Genetic Algorithm (SGA).

The paper is structured as follows: Section 2 provides an overview of related works, and Section 3 contains a description of the WSN with all the hypothesis and parameters adopted. Section 4 provides a brief description of the optimization algorithm. Section 5 reports the results of the optimization and the comparison. Finally, in Section 6, some conclusions are drawn.

2. Related Works

Evolutionary optimization algorithms have been widely applied to optimization problems related to wireless sensor networks. The task assignments of WSN essentially aims to save energy, reducing the expenditure among the allocations and prolonging the network lifetime. Some authors had already proposed energy-efficiency approaches based on a mix of entropy theory and evolutionary computation theory [20].

Moreover, the same problem has been also addressed by means of several evolutionary algorithms: for example, in [21], task scheduling in heterogeneous distributed systems was evaluated comparing a multi-objective evolutionary algorithm with a hybrid genetic algorithm. Previously, genetic algorithms had been used in [22] for a similar purpose, performing tests on a Java scheduler with a homogeneous set of processors. Additionally, in [23], the authors proposed three different hybridizations between BBO and DE for the problem of power allocation in sensor networks.

Particle swarm optimization has been widely applied in WSN: the work in [24] provided a wide review of these applications. As concerns the specific problem of task allocation, in [25], the authors applied the PSO to the this problem with a particular concern regarding the network reliability. In [26], the authors proposed the PSO in task optimization for tracking, showing that the heuristic algorithm can compete and win against deterministic approaches to the same problem. More recently, in [27], a logic-based evolutionary algorithm compared to a binary PSO has been applied again to task allocation in WSN.

Social network optimization has been preliminary applied to task allocation in WSN [28], but in this previous work, the scheduling algorithm was only partially considered and no energy constraints were included. Additionally, no comparison with other assessed evolutionary algorithms was proposed in order to validate SNO performance over the task allocation problem. In the present paper, all these aspects have been added, as clarified in the following sections.

3. Description of Wireless Sensor Networks

The analyzed problem in this paper is the tasks' resource allocation in wireless sensor networks [29].

The WSN is a system composed of a set of smart sensor devices, deployed in space, which can sense, process, send, and receive data. The communication adopts a multi-hop scheme for managing the signal-to-noise ratio without drastically compromising the global lifetime. Moreover, this structure can handle specific environmental constraints.

The most important node in the network is the *cluster head*: this node is devoted to transmission and reception of the data. It is in charge of splitting and gathering the data between the network nodes, and it cannot perform any further processing activity.

All the other standard nodes can be operated in four different modes:

- computing mode: this is the phase related to the elaboration of the assigned task;
- Reception (RX) mode: all the nodes can receive information from their neighbors;
- Transmission (TX) mode: when a sensor is in this mode, it is transmitting the information to its *predecessor* in the network;
- idle state: this is when a sensor is not performing any activity; in this state, the energy consumption is drastically reduced.

Thus, the sensors can be only in one of these four states and cannot perform more than one activity at the same time. On the other hand, the cluster head is devoted only to TX and RX operations. Due to this design option, it is equipped with a multi-input and multi-output transmission capability.

The sensor units can differ one from another in terms of processing speed, while the transmission rate is homogeneously distributed among all the nodes. This is justified due to the fact that, while the transmission rate is related to the signal frequency, the units can be different or can be set with different energy levels to improve the network lifetime.

The specific selected application, namely the parallel computing field, imposes some constraints. Firstly, all the tasks require almost the same computational effort; for the sake of simplicity, in this paper, they have been considered equal. Moreover, a process cycle of the entire network consists of the transmission of the tasks from the cluster head to the sensors, the following elaboration, and in the final collection of the processed outputs. The following cycle can start only when all the outputs are collected, as represented in Figure 1.

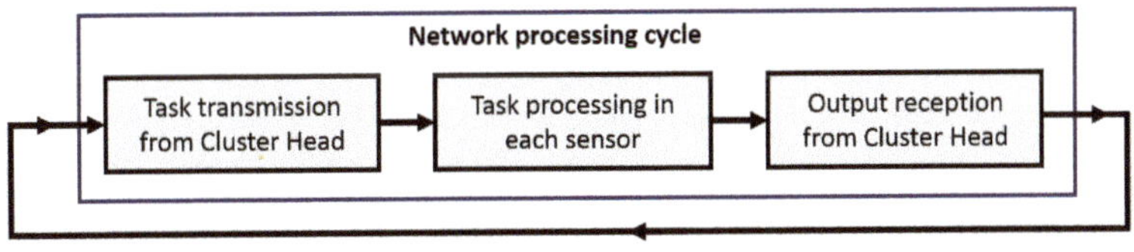

Figure 1. Representation of the network processing cycle: firstly, the information is sent from the cluster head to the sensors, then it is processed, and finally, the outputs are gathered.

The scheduling process of this network is assumed to be in the *first-in-first out* (FIFO) logic. This control scheme has been selected because of its simplicity; even if it is not optimal by itself, it is part of the optimization process, so this sub-optimality can be reduced or eliminated. This logic has the advantages of being simple and can be implemented in the sensors, and it is valid for any kind of job the units should perform. Moreover, this logic is robust to some variation in the elaboration time of the other units.

Within this network, two aspect have been considered: the total cycle time and the network lifetime. The first is the maximum of the time required by the sensors:

$$t_{cycle} = \max_i t_i \tag{1}$$

On the other hand, the network lifetime is the number of cycles before the most stressed node ends its energy.

The total time required by the ith sensor unit is the sum of the time spent in each of the four possible states:

$$t_i = t_{el,i} + t_{RX,i} + t_{TX,i} + t_{w,i} \tag{2}$$

where t_i is the total time required by the ith unit, $t_{el,i}$ is its elaboration time, $t_{RX,i}$ is its receiving time, $t_{TX,i}$ is its transmission time, and $t_{w,i}$ is its idle time.

These times are all a function of the number of allocated tasks to each unit. Firstly, the processing time can be expressed as:

$$t_{el,i} = \frac{N_i \cdot c_{el}}{v_{el,i}} \tag{3}$$

where N_i is the number of allocated tasks to the ith unit, c_{el} is the number of Kilo Clock Cycles (KCC) required for elaborating one task, and $v_{el,i}$ is its elaboration speed in KCC/s.

The time spent in reception mode depends on the network configuration, and it has two contributions: the time required for receiving the tasks from its predecessor and the time for receiving the output of the elaborations from reachable units (in graph theory, the set of reachable nodes is the set of nodes connected directly or indirectly to the analyzed node):

$$t_{RX,i} = \sum_{j=1}^{U_i} \frac{N_j \cdot c_{in}}{v_{RX}} + \sum_{k=1}^{U_i} \frac{N_k \cdot c_{out}}{v_{RX}} = \sum_{j=1}^{U_i} \frac{N_j \cdot (c_{in} + c_{out})}{v_{RX}} \tag{4}$$

where U_i is the number of reachable units from the ith sensor, c_{in} is the information amount sent from the cluster head to the nodes to assign a task, c_{out} is the output number of bits for each assigned task, and v_{RX} is the reception speed.

The transmission time can be calculated with a very similar logic as the reception time. Nonetheless, it can be calculated in an easier way as a function of the reception time: in fact, it is the time for retransmitting the received information and to transmit the output of the task processed by the unit:

$$t_{TX,i} = t_{RX,i} \cdot \frac{v_{RX}}{v_{TX}} + \frac{N_i}{v_{TX}} \tag{5}$$

where v_{TX} is the transmission speed. The transmission and the reception speed can be different because the transmission requires also the amplification of the signal.

The waiting time evaluation depends on the specific schedule of the activities and thus cannot be expressed as the other times, but it can be calculated at each simulation run of the network.

The scheduling is based on the following rules. The information is transmitted from the cluster head to the nodes, giving priority to the nodes with more subsequent nodes; when a node has finished the reception and the hopping of the information, it starts its elaboration; then, it sends back the information to its predecessorif it is in idle mode. Finally, it turns to idle mode for the information.

An example of the scheduling is proposed in Figure 2.

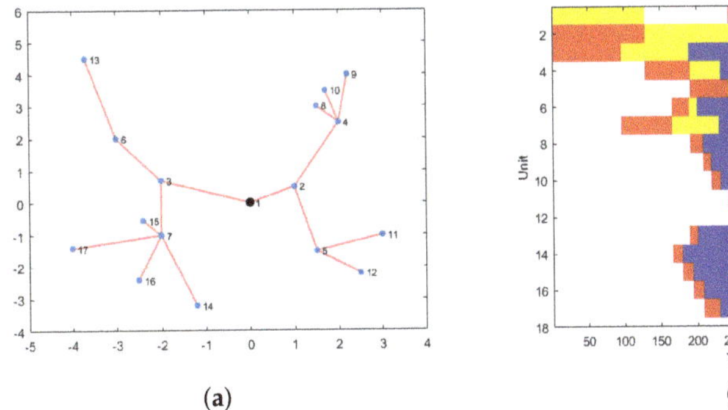

Figure 2. Scheduling example on a simple network. (**a**) Simple network with 17 nodes. (**b**) Scheduling example on one cycle.

As concerns the energy consumption, the following model has been applied. The sensors are fed by batteries with a total capacity of 9 kJ. The energy consumed is composed of four terms:

$$E_i = E_{TRX,i} + E_{amp,i} + E_{el,i} + E_{w,i} \tag{6}$$

The first term is the energy for the communication. It is a function of the total number of bit transmitted and received ($b_{TRX,i}$):

$$E_{TRX,i} = b_{TRX,i} \cdot E_{TRX} \tag{7}$$

where E_{TRX} is the energy required to maintain the communication equipment.

The amplification energy is required to have an acceptable signal-to-noise ratio. The transmission energy required by node i to transmit to node j depends on the number of transmitted bits (b_{ij}) and on the transmission distance (d_{ij}):

$$E_{amp,i,j} = E_{amp} \cdot b_{ij} \cdot d_{ij} \tag{8}$$

The elaboration energy depends on the elaboration time:

$$E_{el,i} = E_{el} \cdot t_{el} \tag{9}$$

Finally, the energy consumed in idle time is considered negligible with respect to the other terms. The network has been implemented in MATLAB; the parameters adopted in the simulation of the network are summarized in Table 1.

Table 1. List of network parameters. KCC, Kilo Clock Cycles.

Parameter	Symbol	Value	Parameter	Symbol	Value
Communication speed	v_{TRX}	25 Mbps	Communication energy	E_{TRX}	5 pJ/b
Stored energy	E_s	9 kJ	Amplification energy	E_{amp}	0.01 pJ/b
Elaboration speed	v_{el}	[30–100] KCC/s	Elaboration energy	E_{el}	30 mW
Input bytes for task	c_{in}	500 byte	Output bytes for tasks	c_{out}	700 bytes
Computational effort for tasks	c_{el}	150 KCC/s	Total number of tasks	N_{tot}	10,000

4. Social Network Optimization

The optimization algorithm used in this paper is Social Network Optimization (SNO). It is a novel population-based algorithm that takes its inspiration from the information-sharing process in online social networks [30]. The algorithm has already been applied to other engineering problems in which its effectiveness has been proven [17].

The performances of this algorithm are very good in many engineering application problems, both in terms of convergence speed and reliability of the solution, and this makes the algorithm very suitable for facing the problem complexity in an affordable time [31].

The basic element of SNO is the social network itself ($\mathbb{S}$). It is a network of users ($\{u_u\}$) that communicates by means of posts ($\{p_s\}$):

$$\mathbb{S} = \{\{u_u\}, \{p_s\}\} \tag{10}$$

Each user is characterized by a set of *opinions* ($\mathbf{o}_u$) (the bold means that it is a vector), its personal growth ($\mathbf{c}_u$), a list of friends ($\mathbb{L}_u$), and a reputation value for the other users of the social network (r_{uv}):

$$u_u = \{\mathbf{o}_u, \mathbf{c}_u, \mathbb{L}_u, \{r_{uv}\}\} \tag{11}$$

The interaction between users takes place among two preferential paths: the first one is a friend network, where the connections between users are very strong and reciprocal, and a trust network, where the connections are weaker and monodirectional. Among these paths, the information is exchanged by means of posts: the information content is presented by a *status* containing the *opinions* of the user on the discussion *topics*.

The topic represents, out of the social metaphor, the possible design variables, while the status is the value assigned to them by each post that represents a candidate solution.

Each post contains the status, $\mathbf{s}_s$, the name (the ID) of the user that has posted it (u_s), the time at which it was posted t_s, and a visibility value (v_s), that is, out of the metaphor, the cost value associated with each candidate solution by means of the objective function. Posts with high visibility are more likely to be seen by the other users, and thus can influence more individuals. The post visibility affects the reputation list: in fact, if the visibility of a post of the user v is higher than the average, its reputation r_{uv} grows, while if it is lower than the average, the reputation diminishes.

$$p_s = \{\mathbf{s}_s, u_s, t_s, v_s\} \tag{12}$$

The opinions of a user are related o the corresponding status by means of the *linguistic transposition* of the ideas. In SNO, it is modeled as a random variable with zero mean and a small standard deviation.

$$\mathbf{s}_u = \mathbf{o}_u + \lambda_u \tag{13}$$

Figure 3 is representative of the basic structure of social network optimization and of its relation with a generic optimization problem.

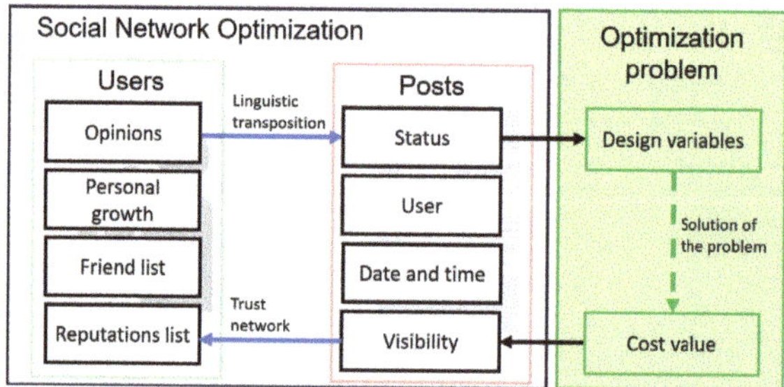

Figure 3. Internal structures of social network optimization and their relation to the optimization problem.

All the above presented structures evolve with time. There are two basic mechanisms in this evolution: the personal growth of individuals and the modification of the friend and trust networks.

The personal growth can be represented by the variation of the opinions with time:

$$c_u(t) = \frac{d\mathbf{o}_u(t)}{dt} \tag{14}$$

The user change $c_u(t)$ is based on the complex contagion model of idea diffusion:

$$c_u(t+1) = \alpha \cdot c_u(t) + \beta \cdot (\mathbf{a}_u(t) - \mathbf{o}_u(t)) \tag{15}$$

where $\mathbf{a}_u(t)$ is the attracting idea created in the interaction process; α and β are two user-defined parameters that represent the inclination of each user to change his/her idea.

The attracting idea is created by means of the friend and trust networks. In fact, each user selects some influencers in these networks by means of a rank selection based on the visibility value. Then, the identified posts are combined with a crossover operator to create the attracting idea. The networks' evolutions are intrinsically different. The friend network evolves by means of the process of recommendation, so a user v has a probability of becoming a friend of a user u that is proportional to the number of common friends. On the other hand, the friendship can be eliminated when the number of common friends becomes low. Furthermore, the trust network is based on the reputation value: all the individuals that for the user u have a reputation higher than a predefined threshold belong to its trust network.

The time evolution of the structures of SNO is described in the flowchart of Figure 4.

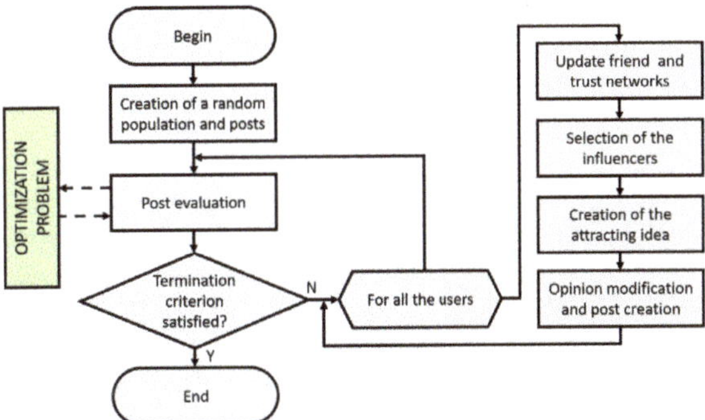

Figure 4. Flowchart of social network optimization.

5. Optimization Problem: Description and Results

The described wireless sensor network represents the engineering problem to which social network optimization has been here applied. In the following, the optimization problem is formalized, and the results are presented.

5.1. Problem Description

The optimization problem is the following one:

$$\min_{T \in \Theta} t_{max} \tag{16}$$

subject to:

$$\sum_{i=1}^{N} T_i = N_{task}$$

$$E_{max} < \frac{E_s}{4000}$$

T is the design variable vector (the number of tasks associated with each sensor), and Θ is the design variable space. The minimization function is the maximum cycle time, i.e., the time required by the slowest sensor in the network.

There are two constraints that were introduced: the first one ensures that all the tasks are managed by the network, while the second one that the network lifetime lasts more than 4000 cycles.

This problem has been codified in SNO by means of a candidate solution represented by a vector of N elements, where N is the number of sensors, and each of them can range from 0–1. The first constraint has been managed with a proper decodification of the candidate solution. In fact, it is possible to process the candidate solution **s** in the following way:

$$\mathbf{T} = \left\lfloor \frac{\mathbf{s}}{|\mathbf{s}|} \cdot N_{tot} \right\rfloor \tag{17}$$

In this way, the produced vector **T** is composed only by an integer. On the other hand, it is not ensured that the sum is equal to N_{tot}. To fix this aspect, the missing task (N_{miss}) are calculated and assigned to the first N_{miss} sensors.

The management of the second constraint has been done with a penalty approach. This means that the actual visibility value for the optimizer is:

$$\mathbf{v} = t_{max} + \left[500 + 10 \cdot (4000 - t_{life}) \right] \cdot H(4000 - t_{life}) \tag{18}$$

where t_{life} is the lifetime of the network expressed in elaboration cycles and $H(\cdot)$ is the Heaviside function defined as:

$$H(x) = \begin{cases} 0, & x < 0 \\ 1, & x \geq 0 \end{cases} \tag{19}$$

This penalty definition has been selected because it ensures that a solution that satisfies the constraints has a cost value greater than a good solution. It has been decided to not further penalize the solutions because they can have good features that could help the convergence process.

In the following, the optimization process has been performed firstly without taking this last constraint into consideration, and then considering it. In both cases, firstly, the results obtained by means of SNO are presented, and then it is compared with other algorithms.

The algorithms adopted for the comparison are the following:

- Biogeography-Based Optimization (BBO), implemented starting from [32] and modified to improve the exploration;
- Differential Evolutionary (DE) [33];
- Genetic Algorithm (GA) [34], implemented for the real value objective function, with single-point crossover and non-linear rank-based selection;
- Particle Swarm Optimization (PSO) [35], with variable inertia and velocity constraints;
- Stud-Genetic Algorithm (SGA) [36], an effective variation of GA in which one of the parents is always the best individual in the population and the second one is selected with a rank-based selection.

For all these algorithms, the population was set to 20 individuals (this value was obtained from a parametric analysis performed on standard benchmarks [37]), and the termination criterion was set to 5000 objective function evaluations. In all the tests, 100 independent trials were performed to have statistical reliability of the results.

5.2. Results of Unconstrained Optimization

The first optimization set of trials was done without the constraint on the minimum lifetime. Firstly, the results of SNO are presented, and then, they are compared with the results of the other algorithms.

5.2.1. Social Network Optimization Results

SNO has been used to find an optimal solution of the WSN task allocation problem. Figure 5 shows the convergence curves of the 100 independent trials. The small plot is a zoom of the convergence curves limiting the cost value to 1600.

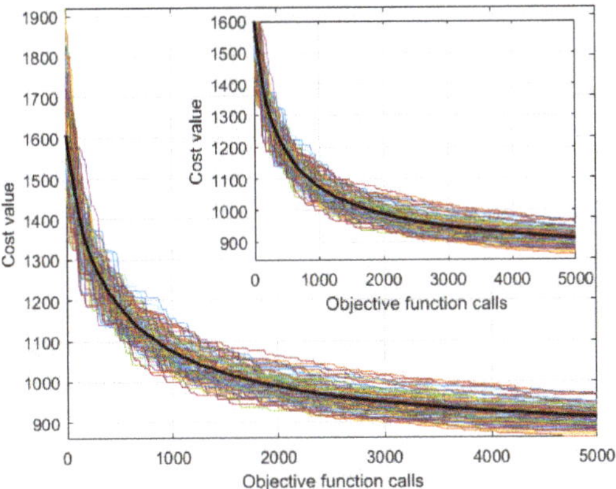

Figure 5. Convergence curves of SNO on 100 independent trials for the unconstrained problem.

The convergence curves have a very similar behavior, showing the robustness of the algorithm. All the trials were able to find solutions with the cost value below 1000.

It is possible to represent the obtained solutions in a plane in which the horizontal axis is the maximum cycle time (in milliseconds) and the vertical axis is lifetime (in elaboration cycles). Figure 6 shows the 100 optimal values obtained by SNO. The three green dots are the Pareto front. In this case, it is possible to notice that many solutions have a lifetime below 4000 cycles. All these solutions will be discarded in the constrained optimization.

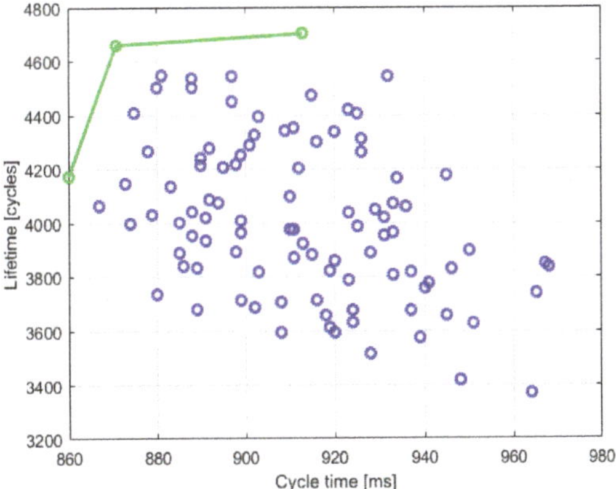

Figure 6. Optimal solutions found by SNO in the cycle time–lifetime plane. The green represents the Pareto front.

The optimal scheduling of one elaboration cycle is represented in Figure 7a. The red color represents the receiving mode, the blue the elaboration status, and the yellow the transmitting mode. White color is the idle status.

Figure 7b shows the energy content of all the sensors during the first five elaboration cycles. The black vertical lines show the end of each cycle. The color is representative of the energy content: blue means full of charge.

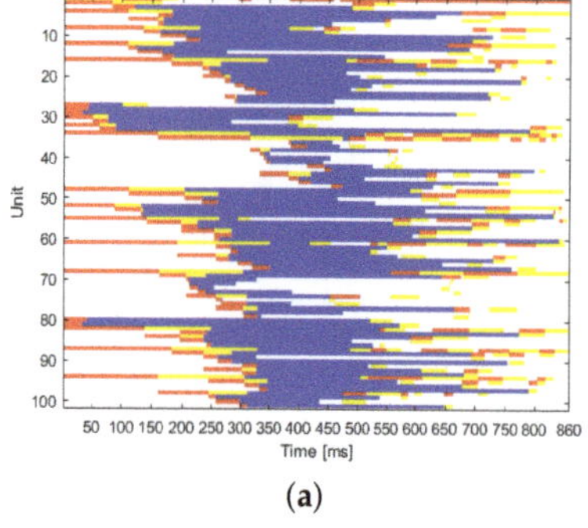

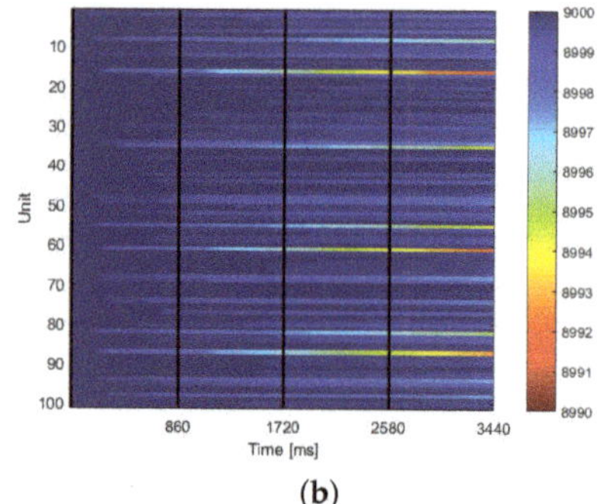

(a)

(b)

Figure 7. Best solution: scheduling and energy level. (**a**) Scheduling of one elaboration cycle. Red represents the receiving mode, blue the elaboration status, and yellow the transmitting mode. (**b**) Energy level for each sensor in the first five elaboration cycles.

5.2.2. Comparison between SNO and Other Algorithms

In this section, the comparison between SNO and the other optimization algorithms is presented.

Table 2 shows the results of the six algorithms: the mean value is the average optimal result obtained in the 100 independent trials; the minimum value is the best solution; and the last column is the results of a *t*-test with the significance level at 5%: if SNO is better, the value of the *t*-test is "+"; if the two algorithms are equal, it is "0", or if SNO is the worst, the value is "−".

It is possible to see that SNO outperformed four out of the five algorithms and that only Stud-GA had similar performances. In particular, Stud-GA is able to find a better best solution, while the average results of the two algorithms are almost the same.

Table 2. Results of the algorithms.

Algorithm	Mean Value	Minimum Value	*t*-Test
BBO	966.41	917	+
DE	938.04	887	+
GA	998.66	915	+
PSO	1120.04	982	+
SGA	911.57	844	0
SNO	911.73	860	

Figure 8a shows a comparison in terms of convergence curves: the continuous line is the average convergence, while the dashed line is the best trial. Figure 8b shows a comparison of the Pareto front: in this figure, it is possible to notice that the results of SNO are much better than the results of the other algorithms.

The convergence of Stud-GA is slower than the one of SNO at the beginning of the optimization process, while it is able to reach a comparable result at the end. On the other hand, BBO has a very good initial convergence, but then the convergence rate drastically diminishes.

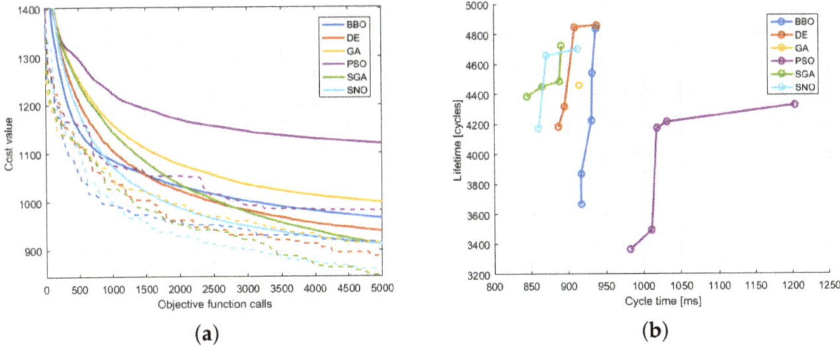

Figure 8. Comparison of the algorithms in terms of convergence curves and Pareto fronts. (**a**) Comparison of convergence curves. (**b**) Comparison of Pareto fronts.

5.3. Results of Constrained Optimization

Here, the results of the constrained optimization are presented. As has been done before, firstly, the results of SNO and then the comparison are presented.

5.3.1. Social Network Optimization Results

Figure 9 shows the convergence curves of the 100 independent trials. Here, the zoom is important to see the good convergence of the trials and the low standard deviation. Also in this case, it is possible to notice that all the results had a cost value below 1000.

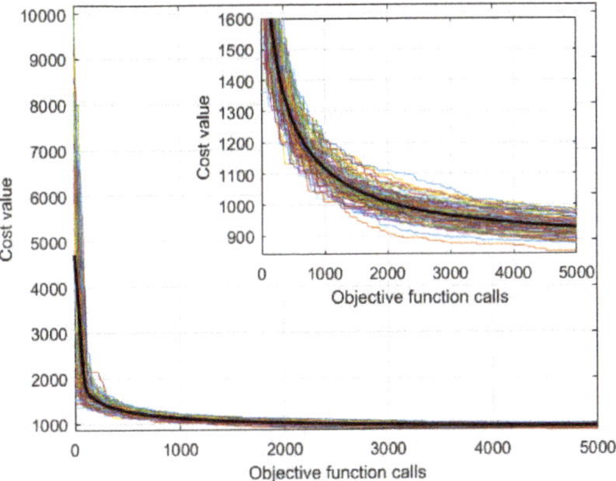

Figure 9. Convergence curves of SNO on 100 independent trials.

With respect to the curves of the unconstrained optimization, it is possible to see that the initial values have a very high cost. This is due to the fact that the solutions found violate the feasibility constraint. After a few iterations, the algorithm is able to bring all the solutions within the feasibility limits, and then, the convergence rate becomes similar to the unconstrained optimization.

As done previously, it is possible to represent the obtained solutions as the maximum cycle time–lifetime plane. Figure 10 shows the 100 optimal values obtained by SNO. The four green dots are the Pareto front.

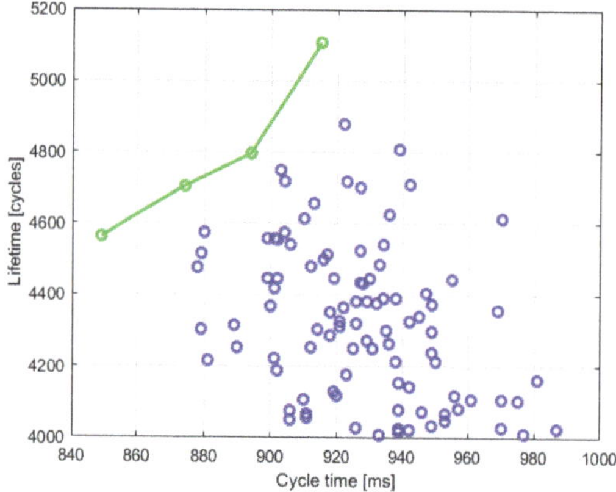

Figure 10. Optimal solutions found by SNO in the cycle time–lifetime plane. The green represents the Pareto front.

The optimal scheduling of one elaboration cycle is represented in Figure 11a, while Figure 11b shows the energy levels of the sensors in the first five elaboration cycles.

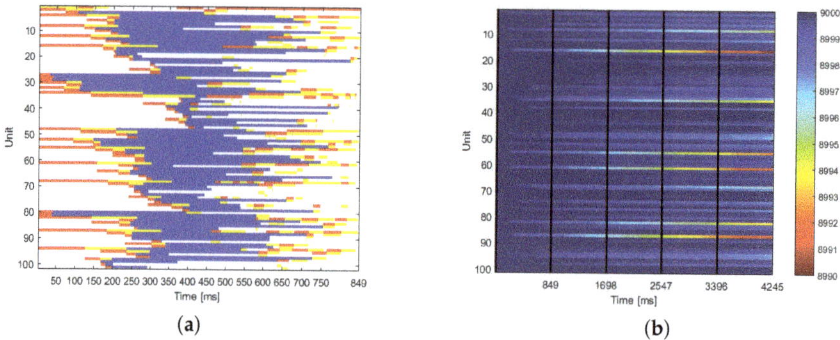

(a)   **(b)**

Figure 11. Best solution: scheduling and energy level. (**a**) Scheduling of one elaboration cycle. Red represents the receiving mode, blue the elaboration status, and yellow the transmitting mode. (**b**) Energy level for each sensor in the first five elaboration cycles.

Comparing these results with the ones of the unconstrained optimization, it is possible to see that, here, the cycle time is lower. Comparing the two schedules, the solutions are very similar. On the other hand, the energy consumption is more distributed among the sensors.

5.3.2. Comparison between SNO and Other Algorithms

In this section, the comparison between SNO and the other optimization algorithms is presented.

Table 3 shows the results of the six algorithm: the mean value is the average optimal result obtained in the 100 independent trials; the minimum value is the best solution; and the last column is the results of a *t*-test with the significance level at 5%.

Table 3. Results of the algorithms.

Algorithm	Mean Value	Minimum Value	*t*-Test
BBO	996.96	929	+
DE	946.52	905	+
GA	1038.32	971	+
PSO	1178.9	1023	+
SGA	924.78	867	0
SNO	925.57	849	

Figure 12a shows a comparison in terms of convergence curves: the continuous line is the average convergence, while the dashed line is the best trial. Figure 12b shows a comparison of the Pareto front: in this figure, it is possible to notice that the results of SNO are much better than the ones of the other algorithms.

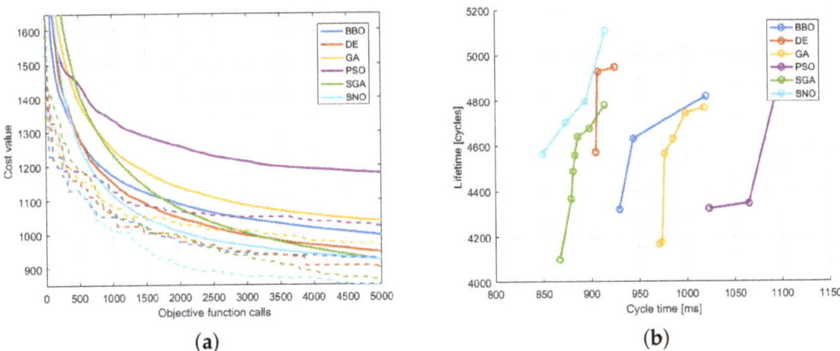

Figure 12. Comparison of the algorithms in terms of convergence curves and Pareto fronts. (**a**) Comparison of convergence curves. (**b**) Comparison of Pareto fronts.

6. Conclusions

In this paper, the task allocation problem in WSN has been faced: the integrated use of evolutionary techniques can optimize and enhance these systems, both in terms of energy efficiency and computationally efficiency.

The optimization algorithm used in this paper is a recently-developed algorithm called social network optimization: this algorithm has been applied in two different problems. The first one is represented by an unconstrained optimization, while the second is a constrained problem.

The obtained results have been compared with other well-established algorithms. The results show that SNO is able to obtain very good results if compared with the other algorithms. In particular, the convergence of SNO is more effective, especially at dealing with the constrained problem.

Future developments of this work can be focused more on the robustness of the WSN communication process: in fact, it is possible to take also into account failures in the information exchange between sensors. This can have an impact on both the total cycle time and the final energy consumption.

Author Contributions: Conceptualization, A.N., F.G., and M.M.; data curation, F.G.; investigation, F.G. and M.M.; methodology, A.N.; software, A.N.; supervision, R.Z.; validation, A.N. and M.M.; writing, original draft, A.N.; writing, review and editing, A.N., F.G., M.M., and R.Z.

Funding: This research received no external funding.

Conflicts of Interest: The authors declare no conflict of interest.

Abbreviations

The following abbreviations are used in this manuscript:

WSN   Wireless Sensor Network
KCC   Kilo Clock Cycles
SNO   Social Network Optimization
PSO   Particle Swarm Optimization
GA   Genetic Algorithm
BBO   Biogeography Based Optimization
SGA   Stud-Genetic Algorithm

References

1. Yick, J.; Mukherjee, B.; Ghosal, D. Wireless sensor network survey. *Comput. Netw.* **2008**, *52*, 2292–2330. [CrossRef]
2. Karakaya, M.; Qi, H. Distributed target localization using a progressive certainty map in visual sensor networks. *Ad Hoc Netw.* **2011**, *9*, 576–590. [CrossRef]
3. Fasolo, E.; Rossi, M.; Widmer, J.; Zorzi, M. In-network aggregation techniques for wireless sensor networks: A survey. *IEEE Wirel. Commun.* **2007**, *14*, 70–87. [CrossRef]
4. Tian, Y.; Ekici, E. Cross-layer collaborative in-network processing in multihop wireless sensor networks. *IEEE Trans. Mobile Comput.* **2007**, *6*, 297–310. [CrossRef]
5. Vorobyov, S.A.; Cui, S.; Eldar, Y.C.; Ma, W.K.; Utschick, W. Optimization Techniques in Wireless Communications. *EURASIP J. Wirel. Commun. Netw.* **2009**, *2009*, 567416. [CrossRef]
6. Hu, X.M.; Zhang, J.; Yu, Y.; Chung, H.S.H.; Li, Y.L.; Shi, Y.H.; Luo, X.N. Hybrid genetic algorithm using a forward encoding scheme for lifetime maximization of wireless sensor networks. *IEEE Trans. Evol. Comput.* **2010**, *14*, 766–781. [CrossRef]
7. Caputo, D.; Grimaccia, F.; Mussetta, M.; Zich, R.E. An enhanced GSO technique for wireless sensor networks optimization. In Proceedings of the IEEE Congress on Evolutionary Computation (IEEE World Congress on Computational Intelligence), Hong Kong, China, 1–6 June 2008; pp. 4074–4079.
8. Tian, J.; Gao, M.; Ge, G. Wireless sensor network node optimal coverage based on improved genetic algorithm and binary ant colony algorithm. *EURASIP J. Wirel. Commun. Netw.* **2016**, *2016*, 104. [CrossRef]
9. Mann, R.P.; Namuduri, K.R.; Pendse, R. Energy-Aware Routing Protocol for Ad Hoc Wireless Sensor Networks. *EURASIP J. Wirel. Commun. Netw.* **2005**, *2005*, 501875. [CrossRef]
10. Caputo, D.; Grimaccia, F.; Mussetta, M.; Zich, R.E. Genetical swarm optimization of multihop routes in wireless sensor networks. *Appl. Comput. Intell. Soft Comput.* **2010**, *2010*, 523943. [CrossRef]
11. Omidvar, A.; Mohammadi, K. Particle swarm optimization in intelligent routing of delay-tolerant network routing. *EURASIP J. Wirel. Commun. Netw.* **2014**, *2014*, 147. [CrossRef]
12. Rodway, J.; Krömer, P.; Karimi, S.; Musilek, P. Differential evolution optimized fuzzy controller for wireless sensor network energy management. In Proceedings of the 2016 IEEE International Conference on Fuzzy Systems (FUZZ-IEEE), Vancouver, BC, Canada, 24–29 July 2016; pp. 352–358.
13. Lu, Q.; Li, S.; Zhang, W.; Zhang, L. A genetic algorithm-based job scheduling model for big data analytics. *EURASIP J. Wirel. Commun. Netw.* **2016**, *2016*, 152. [CrossRef] [PubMed]
14. Gheitanchi, S.; Ali, F.; Stipidis, E. Particle Swarm Optimization for Adaptive Resource Allocation in Communication Networks. *EURASIP J. Wirel. Commun. Netw.* **2010**, *2010*, 465632. [CrossRef]
15. Baioletti, M.; Milani, A.; Poggioni, V.; Rossi, F. An ACO approach to planning. In *European Conference on Evolutionary Computation in Combinatorial Optimization*; Springer: Cham, Switzerland, 2009; pp. 73–84.
16. Baioletti, M.; Milani, A.; Poggioni, V.; Rossi, F. Ant search strategies for planning optimization. In Proceedings of the Nineteenth International Conference on Automated Planning and Scheduling, Thessaloniki, Greece, 19–23 September 2009.
17. Grimaccia, F.; Gruosso, G.; Mussetta, M.; Niccolai, A.; Zich, R.E. Design of tubular permanent magnet generators for vehicle energy harvesting by means of social network optimization. *IEEE Trans. Ind. Electron.* **2018**, *65*, 1884–1892. [CrossRef]

18. Niccolai, A.; Grimaccia, F.; Mussetta, M.; Zich, R.E. Reflectarray optimization by means of sno and pso. In Proceedings of the 2016 IEEE International Symposium on Antennas and Propagation (APSURSI), Fajardo, Puerto Rico, 26 June–1 July 2016; pp. 781–782.

19. Ogliari, E.; Niccolai, A.; Leva, S.; Zich, R. Computational intelligence techniques applied to the day ahead pv output power forecast: Phann, sno and mixed. *Energies* **2018**, *11*, 1487. [CrossRef]

20. Shen, Y.; Ju, H. Energy-efficient task assignment based on entropy theory and particle swarm optimization algorithm for wireless sensor networks. In Proceedings of the 2011 IEEE/ACM International Conference on Green Computing and Communications, Chengdu, China, 4–5 August 2011; pp. 120–123.

21. Chen, Y.; Li, D.; Ma, P. Implementation of Multi-objective Evolutionary Algorithm for Task Scheduling in Heterogeneous Distributed Systems. *JSW* **2012**, *7*, 1367–1374. [CrossRef]

22. Page, A.J.; Keane, T.M.; Naughton, T.J. Multi-heuristic dynamic task allocation using genetic algorithms in a heterogeneous distributed system. *J. Parallel Distrib. Comput.* **2010**, *70*, 758–766. [CrossRef] [PubMed]

23. Boussaïd, I.; Chatterjee, A.; Siarry, P.; Ahmed-Nacer, M. Hybridizing biogeography-based optimization with differential evolution for optimal power allocation in wireless sensor networks. *IEEE Trans. Veh. Technol.* **2011**, *60*, 2347–2353. [CrossRef]

24. Kulkarni, R.V.; Venayagamoorthy, G.K. Particle swarm optimization in wireless-sensor networks: A brief survey. *IEEE Trans. Syst. Man Cybern. Part C* **2011**, *41*, 262–267. [CrossRef]

25. Guo, W.; Li, J.; Chen, G.; Niu, Y.; Chen, C. A PSO-optimized real-time fault-tolerant task allocation algorithm in wireless sensor networks. *IEEE Trans. Parallel Distrib. Syst.* **2015**, *26*, 3236–3249. [CrossRef]

26. Liu, M.; Huang, D.-P.; Xu, X.-L. Node task allocation based on pso in wsn multi-target tracking. *Adv. Inf. Sci. Serv. Sci.* **2010**, *2*, 13–18.

27. Ferjani, A.A.; Liouane, N.; Kacem, I. Task allocation for wireless sensor network using logic gate-based evolutionary algorithm. In Proceedings of the 2016 International Conference on Control, Decision and Information Technologies (CoDIT), St. Julian's, Malta, 6–8 April 2016; pp. 654–658.

28. Grimaccia, F.; Mussetta, M.; Niccolai, A.; Zich, R.E. Optimal computational distribution of social network optimization in wireless sensor networks. In Proceedings of the 2018 IEEE Congress on Evolutionary Computation (CEC), Rio de Janeiro, Brazil, 8–13 July 2018; pp. 1–7.

29. Wu, X.; Sharif, B.S.; Hinton, O.R. An improved resource allocation scheme for plane cover multiple access using genetic algorithm. *IEEE Trans. Evol. Comput.* **2005**, *9*, 74–81. [CrossRef]

30. Niccolai, A.; Grimaccia, F.; Mussetta, M.; Pirinoli, P.; Bui, V.H.; Zich, R.E. Social network optimization for microwave circuits design. *Prog. Electromagn. Res.* **2015**, *58*, 51–60. [CrossRef]

31. Ruello, M.; Niccolai, A.; Grimaccia, F.; Mussetta, M.; Zich, R.E. Black-hole PSO and SNO for electromagnetic optimization. In Proceedings of the 2014 IEEE Congress on Evolutionary Computation (CEC), Beijing, China, 6–11 July 2014; pp. 1912–1916.

32. Simon, D. Biogeography-based optimization. *IEEE Trans. Evol. Comput.* **2008**, *12*, 702–713. [CrossRef]

33. Storn, R.; Price, K. Differential evolution—A simple and efficient heuristic for global optimization over continuous spaces. *J. Glob. Optim.* **1997**, *11*, 341–359. [CrossRef]

34. Goldberg, D.E.; Holland, J.H. Genetic algorithms and machine learning. *Mach. Learn.* **1988**, *3*, 95–99. [CrossRef]

35. Kennedy, J. Particle swarm optimization. In *Encyclopedia of Machine Learning*; Springer: Berlin/Heidelberg, Germany, 2010; pp. 760–766.

36. Khatib, W.; Fleming, P.J. The stud GA: A mini revolution? In *Parallel Problem Solving from Nature*; Springer: Berlin/Heidelberg, Germany, 1998; pp. 683–691.

37. Li, X.; Engelbrecht, A.; Epitropakis, M.G. *Benchmark Functions for CEC'2013 Special Session and Competition on Niching Methods for Multimodal Function Optimization*; Tech. Rep.; Evolutionary Computation and Machine Learning Group, RMIT University: Melbourne, Australia, 2013.

MDPI
St. Alban-Anlage 66
4052 Basel
Switzerland
Tel. +41 61 683 77 34
Fax +41 61 302 89 18
www.mdpi.com

Mathematics Editorial Office
E-mail: mathematics@mdpi.com
www.mdpi.com/journal/mathematics